ELECTRONIC TEST INSTRUMENTS

Robert A. Witte

Agilent Technologies

PRENTICE HALL PTR
UPPER SADDLE RIVER, NJ 07458
WWW.PHPTR.COM

ISBN 0-13-066830-3

90000

9 790130 668300

A CIP catalog record for this book can be obtained from the Library of Congress.

Editorial/production supervision: *MetroVoice Publishing*
Cover design director: *Jerry Votta*
Cover design: *Talar Agasyan-Boorujy*
Manufacturing manager: *Alexis Heydt-Long*
Buyer: *Maura Zaldivar*
Executive editor: *Jill Harry*
Editorial assistant: *Sarah Hand*
Marketing manager: *Dan De Pasquale*

© 2002 by Prentice Hall PTR
A Division of Pearson Education, Inc.
Upper Saddle River, New Jersey 07458

Prentice Hall books are widely used by corporations and government agencies for training, marketing, and resale.
The publisher offers discounts on this book when ordered in bulk quantities. For more information, contact Corporate Sales Department, Phone: 800-382-3419; FAX: 201-236-7141;
E-mail: corpsales@prenhall.com
Or write: Prentice Hall PTR, Corporate Sales Dept., One Lake Street, Upper Saddle River, NJ 07458.

Company and product names mentioned herein are the trademarks or registered trademarks of their respective owners.

Printed in the United States of America
10   9   8   7   6   5   4   3   2   1

ISBN 0-13-066830-3

Pearson Education LTD.
Pearson Education Australia PTY, Limited
Pearson Education Singapore, Pte. Ltd.
Pearson Education North Asia Ltd.
Pearson Education Canada, Ltd.
Pearson Educación de Mexico, S.A. de C.V.
Pearson Education—Japan
Pearson Education Malaysia, Pte. Ltd.

To my family—Joyce, Sara, and Rachel

CONTENTS

Chapter 10   Circuits for Electronic Measurements   311

PREFACE

This book is for the electrical engineer, technician, or student who understands basic electronics and wants to learn more about electronic measurements and test instruments. To use electronic instruments effectively, it is necessary to understand basic measurement theory and how it relates to practical measurements. Basic measurement theory includes such things as how a voltage waveform relates to its frequency and how an instrument can affect the voltage that it is measuring. In an ideal world, we would not have to know anything about the internal operation of an instrument to use it effectively. Although this ideal situation can be approached, it cannot be obtained completely. (One does not have to know how a gasoline engine works to drive an automobile. However, a driver does need to understand the function of the accelerator and brake pedals.)

To minimize dealing with the internal workings of an instrument, circuit models and conceptual block diagrams are used extensively. Circuit models take a "black box" approach to describing a circuit. In other words, the behavior of a complex circuit or instrument can be described adequately by conceptually replacing it with a much simpler circuit. This circuit model approach reduces the amount of detail that must be remembered and understood. Conceptual block diagrams show just enough of the inner workings of an instrument so that the reader can understand what the instrument is doing, without worrying about the details of how this is accomplished.

In all instrument categories, the traditional analog technologies have been overtaken by digital technology. More precisely, the old analog approach has been replaced by precision analog circuitry that is enhanced by the power of analog-to-digital converters, digital logic, digital signal processing, and measurement algorithms implemented via software. However, a voltage measurement is still a voltage measurement, whether an analog meter or a digital meter is used. Since the measurement is fundamentally the same, this book treats both technologies in a unified manner, emphasizing digital instruments and highlighting the differences between the analog and digital approaches when appropriate.

This book does not attempt to be (nor can it be) a substitute for a well-written instrument operating manual. The reader is not well served by a book that says "push this button, turn this knob" because the definition of the buttons and knobs will undoubtedly change with time. Instead, this book is a reference, which provides the reader with a background in electronic instruments. Variations and improvements in instrument design cause each meter, oscilloscope, or function generator to be somewhat unique. However, they all have in common the fundamental measurement principles covered in this book.

This second edition of the book includes updates to all of the chapters, incorporating recent developments in technology while still remaining focused on the concepts and principles that last over time. The oscilloscope chapters were expanded, with an increased emphasis on digital oscilloscopes. The section on power supplies was expanded into its own chapter.

Chapter 1 covers the basic measurement theory and fundamentals. Chapters 2 through 7 cover the mainstream instruments and applications that the typical user will encounter (meters, signal sources, oscilloscopes, frequency counters, and power supplies). Chapter 8 introduces spectrum analyzer, network analyzers, and RF power meters while Chapter 9 covers logic probes and logic analyzers. Chapter 10 rounds out the book with some important circuit concepts and techniques that enable quality measurements.

My original motivation to write this book was my experience in teaching electrical engineering circuit theory courses. Even students with a good background in electrical theory seem to have trouble relating the textbook concepts to what is observed in the laboratory. The concepts of the loading effect, grounding, and bandwidth are particularly troublesome, so they are emphasized throughout the book.

Acknowledgments

My thanks goes to all of the talented engineers, technicians, professors, and students with whom I have had the privilege of working, learning, and playing (not necessarily in that order). This book reflects the experiences I've shared and the knowledge received from many people over time.

I wish to express my gratitude to the people at Prentice Hall that helped make this book possible, especially my editor, Jill Harry. (Jill, you probably thought I would never finish this project. Thanks for hanging in there with me.) My special thanks goes to my colleagues who provided valuable feedback or other assistance concerning the revision of this book: Virgil Leenerts, Mark Lombardi, Ken Wyatt, Steve Warntjes, David Grieve, and Robb Bong. My special appreciation goes to my spouse, Joyce Witte, for assisting with the editing. I'd also like to thank these people for their help in creating the first edition: James Kahkoska, Jerry Murphy, Bill Spaulding, Scott Stever, Ken Wyatt, and Joyce Witte.

—Bob Witte, February 2002

CHAPTER 1

Measurement Theory

To understand the operation and use of electronic instruments, it is necessary to have a solid background in the electrical theory associated with electronic measurements. Although it is assumed that the reader understands basic electrical principles (voltage, current, Ohm's Law, etc.), these principles will be reviewed here with special emphasis on how the theory relates to electronic measurement. With this approach, the theory is used to lay the groundwork for discussing the use and operation of electronic instruments. Most of the fundamental concepts apply to multiple types of measurements and instruments.

1.1 Electrical Quantities

The purpose of electronic test instruments is to accurately measure electrical parameters. Before attempting to discuss electronic measurements, we will first make sure that the parameters to be measured are understood. Appendix A contains a table of electrical parameters, their units of measure, and standard abbreviations. The standard electrical units can be modified by the use of prefixes (milli, kilo, etc.), which are also covered in Appendix A. Our initial concern is with the measurement of voltage and current (with an emphasis on voltage); later we include other parameters such as capacitance and inductance.

Current (measured in units of amperes and often abbreviated to "amps") is the flow of electrical *charge* (measured in coulombs). The amount of charge flowing is determined by the number of electrons moving past a given point. An electron has a negative charge of 1.602×10^{-19} coulombs, or equivalently, a coulomb of negative charge consists of 6.242×10^{18} electrons. The unit of current (the ampere) is defined as the number of coulombs of charge passing a given point in a second (one ampere equals one coulomb per second). The more charge that moves in a given time, the higher the current. Even though current is usually made up of moving electrons, the standard electrical engineering convention is to consider current to be the flow of positive charge.[1] With this definition, the current is considered to be flowing in the direction opposite of the electron flow (since electrons are negatively charged).

Voltage (measured in units of volts), also referred to as *electromotive force* (EMF) or *electrical potential*, is the electrical force or pressure that causes the charge to move and the current to flow. Voltage is a relative concept, that is, voltage at a given point must be specified relative to some other reference point, which may be the system common or ground point.

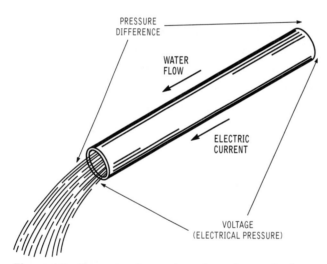

Figure 1.1   The water pipe analogy shows how water flow and pressure difference behave similarly to electrical current and voltage.

An often-used analogy to electrical current is a water pipe with water flowing through it (Figure 1.1). The individual water molecules can be thought of as electrical charge. The amount of water flowing is analogous to electrical current. The water pressure (presumably provided by some sort of external pump) corresponds to electrical pressure or voltage. In this case, the water pressure we are interested in is actually the difference between the two pressures at each end of the pipe. If the pressure (voltage) is the same at both ends of the

---

1.   Current is also sometimes defined as electron flow (i.e., the current flows opposite to the direction used in this book).

pipe, the water flow (current) is zero. On the other hand, if the pressure (voltage) is higher at one end of the pipe, water (current) flows away from the higher pressure end toward the lower pressure end.

Note that while the water flows *through* the pipe, the water pressure is *across* the pipe. In the same way, current flows *through* an electrical device, but voltage (electrical pressure) exists *across* a device (Figure 1.2). This affects the way we connect measuring instruments, depending on whether we are measuring voltage or current. For voltage measurement, the measuring instrument is connected in parallel with the two voltage points (Figure 1.3a). Two points must be specified when referring to a particular voltage and two points are required for a voltage measurement (one of them may be the system common). Measuring voltage at one point only is incorrect. Often, we refer to a voltage at one point when the other point is implied to be the system common or ground point.[2] This is an acceptable practice as long as the assumed second point is made clear.

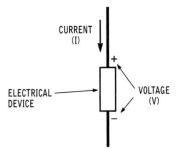

Figure 1.2   Electrical current flows through the device while the voltage exists across the device.

Current, similar to water flow, passes through a device or circuit. When measuring current, the instrument is inserted into the circuit that we are measuring (Figure 1.3b). (There are some exceptions to this, such as current probes, discussed in Chapter 4.) The circuit is broken at the point the current is to be measured and the instrument is inserted in series. This results in the current being measured passing through the measuring instrument. To preserve accuracy for both voltage and current measurements, it is important that the measuring instrument not affect the circuit that is being measured significantly.

---

2.   The concept of an ideal system common or ground can sometimes be misleading as there can be small voltage variations between different locations in the system common.

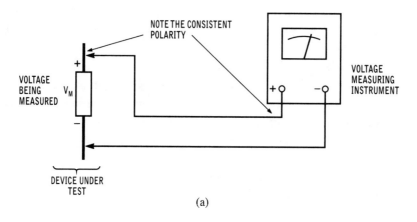

(a)

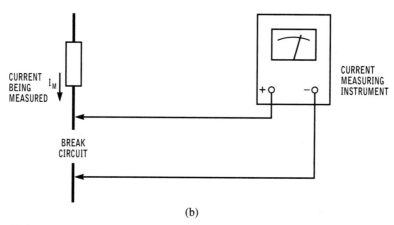

(b)

Figure 1.3   a) Voltage measurements are made at two points. Note that the polarity of the measuring instrument is consistent with the polarity of the voltage being measured. b) Current measurements are made by inserting the measuring instrument into the circuit such that the current flows through the instrument.

1.2 Resistance

A resistor is an electrical device that obeys Ohm's Law:

$$V = I \times R$$

$$I = \frac{V}{R}$$

$$R = \frac{V}{I}$$

Ohm's Law simply states that the current through a resistor is proportional to the voltage across that resistor. Returning to the water pipe analogy, as the pressure (voltage) is increased, the amount of water flow (current) also increases. If the voltage is reduced, the current is reduced. Carrying the analogy further, the resistance is related inversely to the size of the water pipe. The larger the water pipe, the smaller the resistance to water flow. A large water pipe (small resistor) allows a large amount of water (current) to flow for a given pressure (voltage). The name "resistor" is due to the behavior of the device: it resists current. The larger the resistor, the more it resists and the smaller the current (assuming a constant voltage).

1.3 Polarity

Some care must be taken to ensure that voltages and currents calculated or measured have the proper direction associated with them. The standard engineering sign convention (definition of the direction of the current relative to the polarity of the voltage) is shown in Figure 1.4. A voltage source (V) is connected to a resistor (R) and some current (I) flows through the resistor. The direction of I is such that for a positive voltage source, a positive current will leave the voltage source at the positive terminal and enter the resistor at its most positive end. (Remember, current is taken to be the flow of positive charge. The electrons would actually be moving opposite the current.) Figure 1.3 shows the measuring instrument connected up in a manner consistent with the directions in Figure 1.4.

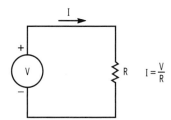

Figure 1.4 Ohm's Law is used to compute the amount of current (I) that will result with voltage (V) and resistance (R).

If the measuring instrument is connected up with the wrong polarity (backwards) when making direct current measurements, the instrument will attempt to measure the proper value, but with the wrong sign (e.g., –5 volts instead of +5 volts). Typically, in digital instruments this is not a problem since a minus sign is just added in front of the reading. In an analog instrument the reading will usually go off the scale and if an electromechanical meter is used it will deflect the meter in the reverse direction, often causing damage. The user should consult the operating manual to understand the limitations of the particular instrument.

Example 1.1

Calculate the amount of voltage across a 5 kΩ resistor if 3 mA of current is flowing through it.

Using Ohm's Law, V = I R = (.003) (5000) = 15 volts

1.4 Direct Current

Direct current (DC) is the simplest form of current. Both current and voltage are constant with respect to time. A plot of DC voltage[3] versus time is shown in Figure 1.5. This plot should seem rather uninformative, since it shows the same voltage for all values of time, but will contrast well with alternating current when it is introduced.

Batteries and DC power supplies produce DC voltage. Batteries are available with a variety of voltage and current ratings. DC power supplies convert AC voltages into DC voltages and are discussed in Chapter 7.

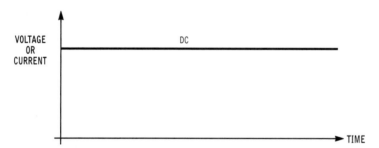

Figure 1.5   DC voltages and currents are constant with respect to time.

1.5 Power

Power is the rate at which energy flows from one circuit to another circuit. For DC voltages and currents, power is simply the voltage multiplied by the current and the unit is the Watt:

$$P = V \times I$$

Using Ohm's Law and some basic math, the relationships in Table 1.1 can be developed. Notice that power depends on both current and voltage. There can be high voltage

---

3.   Even though the term "DC" specifically states direct *current*, it is used to describe both voltage and current. This leads to commonly used, but self-contradictory terminology such as "DC Voltage," which means, literally, "direct current voltage."

but no power if there is no path for the current to flow. Or there could be a large current flowing through a device with zero volts across it, also resulting in no power being received by the device.

Table 1.1 Basic equations for DC voltage, DC current, resistance, and power.

$V = I \times R$	Ohm's Law
$I = \dfrac{V}{R}$	Ohm's Law
$R = \dfrac{V}{I}$	Ohm's Law
$P = V \times I$	Power equation
$P = I^2 R$	Power in Resistor
$P = \dfrac{V^2}{R}$	Power in Resistor

1.6 Alternating Current

Alternating current (AC), as the name implies, does not remain constant with time like direct current, but instead changes direction, or alternates, at some frequency. The most common form of AC is the sine wave, as shown in Figure 1.6. The current or voltage starts out at zero, becomes positive for one half of the cycle, and then passes through zero to become negative for the second half of the cycle. This cycle repeats continuously.

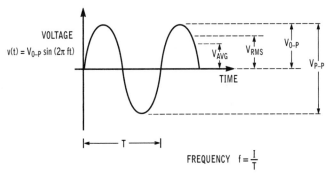

Figure 1.6 The most common form of alternating current is the sine wave. The voltage of the waveform can be described by the RMS value, zero-to-peak value, or the peak-to-peak value.

The sine wave can be described mathematically as a function of time.

$$v(t) = V_{0-P} \sin(2\pi ft)$$

The length of the cycle (in seconds) is called the *period* and is represented by the symbol *T*. The *frequency*, or *f*, is the reciprocal of the period and is measured in units of Hertz, which is equivalent to cycles per second.

$$f = \frac{1}{T}$$

The frequency indicates how many cycles the sine wave completes in one second. For example, the standard AC power line voltage in the United States has a frequency of 60 Hz, which means that the voltage goes through 60 complete cycles in one second. The period of a 60 Hz sine wave is T = 1/f = 1/60 = 0.0167 seconds.

Sometimes the sine wave equation is presented in the following form

$$v(t) = V_{0-P} \sin(\omega t)$$

where ω is the *radian frequency*, with units of radians/second.

By comparing the two equations for the sinusoidal voltage, we can see that

$$\omega = 2\pi f$$

and

$$f = \frac{\omega}{2\pi}$$

Since the sine wave voltage is not constant with time, it is not immediately obvious how to describe it. Sometimes the voltage is positive, sometimes it is negative, and twice every cycle it is zero. This problem does not exist with DC, since it is always a constant value. Figure 1.6 shows four different ways of referring to AC voltage. The *zero-to-peak value* (V_{0-P}) is simply the maximum voltage that the sine wave reaches. (This value is often just called the *peak value*.) Similarly, the *peak-to-peak value* (V_{P-P}) is measured from the maximum positive voltage to the most negative voltage. For a sine wave, V_{P-P} is always twice V_{0-P}.

1.7 RMS Value

Another way of referring to AC voltage is the *RMS value* (V_{RMS}). RMS is an abbreviation for *root-mean-square*, which indicates the mathematics behind calculating the value of an arbitrary waveform. To calculate the RMS value of a waveform, the waveform is first squared at every point. Then, the average or mean value of this squared waveform is found. Finally, the square root of the mean value is taken to produce the RMS (root of the mean of the square) value.

Mathematically, the RMS value of a waveform can be expressed as

$$V_{RMS} = \sqrt{\frac{1}{T} \int_{t_0}^{(t_0 + T)} v^2(t)\,dt}$$

Determining the RMS value from the zero-to-peak value (or vice versa) can be difficult due to the complexity of the RMS operation. However, for a sine wave the relationship is simple. (See Appendix B for a detailed analysis.)

$$V_{RMS} = \frac{1}{\sqrt{2}} V_{0-P} = 0.707 V_{0-P} \qquad \text{(sine wave)}$$

This relationship is valid only for a sine wave and does *not* hold for other waveforms.

1.8 Average Value

Finally, AC voltage is sometimes defined using an average value. Strictly speaking, the average value of a sine wave is zero because the waveform is positive for one half cycle and is negative for the other half. Since the two halves are symmetrical, they cancel out when they are averaged together. So, on the average, the waveform voltage is zero.

Another interpretation of average value is to assume that the waveform has been full-wave rectified. Mathematically this means that the absolute value of the waveform is used (i.e., the negative portion of the cycle has been treated as being positive).

$$V_{AVG} = \frac{1}{T} \int_{t_0}^{(t_0 + T)} |v(t)|\,dt$$

This corresponds to the measurement method that some instruments use to handle AC waveforms, so this is the method that will be considered here. Unless otherwise indicated, V_{AVG} will mean the full-wave rectified average value. These averaging steps have been shown in Figure 1.7. Figure 1.7a shows a sine wave that is to be full-wave rectified. Figure 1.7b shows the resulting full-wave rectified sine wave. Whenever the original waveform becomes negative, full-wave rectification changes the sign and converts the voltage into a positive waveform with the same amplitude. Graphically, this can be described as folding the negative half of the waveform up onto the positive half, resulting in a humped sort of waveform. Now the average value can be determined and is plotted in Figure 1.7b. The relationship between V_{AVG} and V_{0-P} depends on the shape of the waveform. For a sine wave[4]

$$V_{AVG} = \frac{2}{\pi} V_{0-P} = 0.637 V_{0-P} \qquad \text{(sine wave)}$$

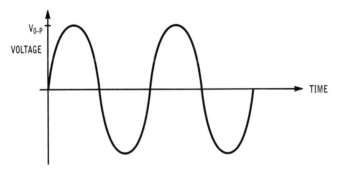

(a) The original sine wave.

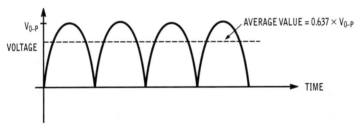

(b) The full-wave rectified version of the sine wave.

Figure 1.7   The operations involved in finding the full-wave average value of a sine wave. a) The original sine wave. b) The full-wave rectified version of the sine wave. The negative half cycle is folded up to become positive. The resulting waveform is then averaged.

The details of the analysis are shown in Appendix B.

1.9 Crest Factor

The ratio of the zero-to-peak value to the RMS value of the waveform is known as the *crest factor*. Crest factor is a measure of how high the waveform peaks, relative to its RMS value. The crest factor of a waveform is important in some measuring instruments. Wave-

---

4. $\pi \approx 3.14159$.

forms with very high crest factors require the measuring instrument to tolerate very large peak voltages while simultaneously measuring the much smaller RMS value.

$$Crest\ factor\ =\ \frac{V_{0-P}}{V_{RMS}}$$

Example 1.2

What is the crest factor of a sine wave?
For a sine wave, $V_{RMS} = 0.707\ V_{0-P}$ so the crest factor is $1/0.707 = 1.414$. The sine wave has a relatively low crest factor. Its zero-to-peak value is not that much greater than its RMS value.

The *peak-to-average ratio*, also known as *average crest factor*, is similar to crest factor except that the average value of the waveform is used in the denominator of the ratio. It also is a measure of how high the peaks of the waveform are compared to its average value.

Although the preceding discussion has referred to AC voltages, the same concepts apply to AC currents. That is, AC currents can be described by their zero-to-peak, peak-to-peak, RMS, and average values.

1.10 Phase

The voltage specifies the amplitude or height of the sine wave; and the frequency or the period specifies how often the sine wave completes a cycle. But two sine waves of the same frequency may not cross zero at the same time. Therefore, the *phase* of the sine wave is used to define its position on the time axis. The most common unit of phase is the degree, with one cycle of a sine wave divided up into 360 degrees of phase. Phase may also be expressed in units of radians, with one cycle corresponding to 2π radians.

The mathematical definition of the sine wave can be modified to include a phase term.

$$v(t)\ =\ V_{0-P}\sin(2\pi ft + \theta)$$

As written, the equation implies that phase is absolute. That is, there is some instant in time when $t = 0$—the reference for the phase angle. In practice, there usually is no such universal time and phase is a relative concept. In other words, we can refer to the phase between two sine waves, but not the phase of a single, isolated sine wave (unless some other time reference is supplied).

For example, the two sine waves in Figure 1.8a are separated by one-fourth of a cycle (they reach their maximum values one-fourth of a cycle apart). Since one cycle equals 360 degrees, the two sine waves have a phase difference of 90 degrees. To be more precise, the

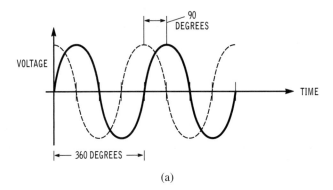

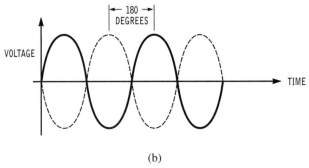

Figure 1.8 The phase of a sine wave defines its relative position on the time axis. a) The phase between the two sine waves is 90 degrees. b) The two sine waves are 180 degrees out of phase.

second sine wave is 90 degrees behind the first or, equivalently, the second sine wave has a phase of –90 degrees relative to the first sine wave. (The first sine wave has a +90 degrees phase relative to the second sine wave.) It is also correct to say that the first sine wave *leads* the second one by 90 degrees or that the second sine wave *lags* the first one by 90 degrees. All of these statements specify the same phase relationship.

Figure 1.8b shows two sine waves that are one half cycle (180 degrees) apart. This is the special case where one sine wave is the negative of the other. The phase relationship between two sine waves simply defines how far one sine wave is shifted with respect to the other. When a sine wave is shifted by 360 degrees, it is shifted a complete cycle and is indistinguishable from the original waveform. Because of this, phase is usually specified over a 360-degree range, typically –180 degrees to +180 degrees.

1.11   AC Power

The concept of power gets more complicated when AC waveforms are considered. The *instantaneous power* is defined by multiplying the instantaneous voltage and instantaneous current:

$$p(t) = v(t) \cdot i(t)$$

In many cases, we are interested in the *average power* being transferred. The average power is determined by taking the instantaneous power and averaging it over a period of the waveform.

$$P = \frac{1}{T}\int_0^T p(t)dt = \frac{1}{T}\int_0^T v(t) \cdot i(t)dt$$

The average power dissipated by a resistor with an AC voltage across it is given by:

$$P = V_{RMS}\,I_{RMS} = \frac{V_{RMS}^2}{R} = I_{RMS}^2 \cdot R$$

This relationship holds for any waveform as long as the RMS value of the voltage and current are used. Note that these equations have the same form as the DC case, which is one of the reasons for using RMS values. The RMS value is often called the *effective value*, since an AC voltage with a given RMS value has the same effect (in terms of power) that a DC voltage with that same value. (A 10-volt RMS AC voltage and a 10-volt DC voltage both supply the same power, 20 watts, to a 5 Ω resistor.) In addition, two AC waveforms that have the same RMS value will cause the same power to be delivered to a resistor. This is *not* true for other voltage descriptions such as zero-to-peak and peak-to-peak. Thus, RMS is the great equalizer with respect to power.

Example 1.3

The standard line voltage in the United States is approximately 120 volts RMS. What are the zero-to-peak, peak-to-peak, and full-wave rectified average voltages? How much power is supplied to a 200 Ω resistor connected across the line?

$$V_{RMS} = 0.707\,V_{0-P}$$
$$\text{so } V_{0-P} = V_{RMS}/0.707 = 120/0.707 = 169.7 \text{ volts}$$
$$V_{P-P} = 2\,V_{0-P} = 2\,(169.7) = 339.4 \text{ volts}$$
$$V_{AVG} = 0.637\,V_{0-P} = 0.637 \times 169.7 = 108.1 \text{ volts}$$

$$P = V_{RMS}^2/R = 120^2/200 = 72 \text{ watts}$$

1.12   Nonsinusoidal Waveforms

There are other AC voltage and current waveforms besides sine waves that are commonly used in electronic systems. Some of the more common waveforms are shown in Table 1.2. Note that the values of V_{RMS} and V_{AVG} (relative to V_{0-P}) are unique for each waveform. The first three waveforms are symmetrical about the horizontal axis, but the half-sine wave and the pulse train are always positive. All of these waveforms are *periodic* because they repeat the same cycle or period continuously.

An example will help to emphasize the utility of RMS voltages when dealing with power in different waveforms.

Example 1.4

A sine wave voltage and a triangle wave voltage are each connected across two separate 300 Ω resistors. If both waveforms deliver 2 watts (average power) to their respective resistors, what are the RMS and zero-to-peak voltages of each waveform?

The two waveforms deliver the same power to identical resistors, so their RMS voltages must be the same. (This is *not* true of their zero-to-peak values.)

$$P = \frac{V_{RMS}^2}{R}$$

$$V_{RMS} = \sqrt{P \cdot R} = \sqrt{2 \cdot 300} = 24.5 \text{ volts RMS}$$

From Table 1.2, for the sine wave

$$V_{RMS} = 0.707 V_{0-P}$$

$$V_{0-P} = \frac{V_{RMS}}{0.707} = \frac{24.5}{0.707} = 34.7 \text{ volts}$$

For the triangle wave

$$V_{RMS} = 0.577 V_{0-P}$$

$$V_{0-P} = V_{RMS}/0.577 = 24.5/0.577 = 42.5 \text{ volts}$$

So for the triangle wave to supply the same average power to a resistor, it must reach a higher peak voltage than the sine wave.

Table 1.2   Table of waveforms with peak-to-peak voltage (V_{P-P}), RMS voltage (V_{RMS}), zero-to-peak voltage (V_{0-P}), and crest factor for each waveform. V_{AVG} is the full-wave rectified average value of the waveform.

	Waveform	V_{P-P}	V_{RMS}	V_{AVG}	Crest Factor
V_{0-P} (sine wave graph)	Sine Wave	$2\,V_{0-P}$	$\frac{1}{\sqrt{2}}V_{0-P}$ or $0.707V_{0-P}$	$\frac{2}{\pi}V_{0-P}$ or $0.637V_{0-P}$	$\sqrt{2}$ or 1.414
V_{0-P} (square wave graph)	Square Wave	$2\,V_{0-P}$	V_{0-P}	V_{0-P}	1
V_{0-P} (triangle wave graph)	Triangle Wave	$2\,V_{0-P}$	$\frac{1}{\sqrt{3}}V_{0-P}$ or $0.577V_{0-P}$	$\frac{1}{2}V_{0-P}$	$\sqrt{3}$ or 1.732
V_{0-P} (half sine wave graph)	Half Sine Wave	V_{0-P}	$\frac{1}{2}V_{0-P}$	$\frac{1}{\pi}V_{0-P}$ or $0.318V_{0-P}$	2
V_{0-P} (pulse train graph)	Pulse Train	V_{0-P}	$\sqrt{\frac{\tau}{T}}V_{0-P}$	$\frac{\tau}{T}V_{0-P}$	$\sqrt{\frac{T}{t}}$

1.13 Harmonics

Periodic waveforms, except for absolutely pure sine waves, contain frequencies called harmonics. Harmonic frequencies are integer multiples of the original or *fundamental* frequency.

$$f_n = n \cdot f_{fundamental}$$

For example, a nonsinusoidal waveform with a fundamental frequency of 1 kHz has harmonics at 2 kHz, 3 kHz, 4 kHz, and so on. There can be any number of harmonics, out to infinity, but usually there is a practical limitation on how many need be considered. Each harmonic may have its own unique phase relative to the fundamental.

Harmonics are present because periodic waveforms, regardless of shape, can be broken down mathematically into a series of sine waves. However, this behavior is more than just mathematics. It is as if the physical world regards the sine wave as the purest, most simple sort of waveform with all other periodic waveforms being made up of collections of sine waves. The result is that a periodic waveform (such as a square wave) is exactly equivalent to a series of sine waves.

1.14 Square Wave

Consider the square wave shown in Figure 1.9a. Square waves are made up of the fundamental frequency plus an infinite number of odd harmonics.[5] In theory, it takes every one of those infinite numbers of harmonics to create a true square wave. In practice, the fundamental and several harmonics will approximate a square wave. In Figure 1.9b, the fundamental, third harmonic, and fifth harmonic are plotted. (Remember, the even harmonics of a square wave are zero.) Note that the amplitude of each higher harmonic is less than the previous one, so the highest harmonics may be small enough to be ignored. When these harmonics are added up, they produce a waveform that resembles a square wave. Figure 1.9c shows the waveform that results from combining just the fundamental and the third harmonic. Already the waveform starts to look somewhat like a square wave (well, a least a little bit). The fundamental plus the third and fifth harmonics is shown in Figure 1.9d. It is a little more like a square wave. Figures 1.9e and 1.9f each add another odd harmonic to our square wave approximation and each one resembles a square wave more closely than the previous waveform. If all of the infinite number of harmonics were included, the resulting waveform would be a perfect square wave. So the quality of the square wave is limited by the number of harmonics present.

The amplitude of each harmonic must be just the right value for the resulting wave to be a square wave. In addition, the phase relationships between the harmonics must also be correct. If the harmonics are delayed in time by unequal amounts, the square wave will take on a distorted look even though the amplitudes of the harmonics may be correct. This phenomenon is used to advantage in square wave testing of amplifiers, as discussed in Chapter 5. It is theoretically possible to construct a square wave electronically by connecting a large number of sine wave generators together such that each one contributes the fundamental frequency or a harmonic with just the right amplitude. In practice, this may be difficult because the frequency and phase of each oscillator must also be precisely controlled.

So far, waveforms have been characterized using a voltage versus time plot, which is known as the *time domain* representation. Another way of describing the same waveform is with frequency on the horizontal axis and voltage on the vertical axis. This is known as the

---

5.   For a mathematical derivation of the frequency content of a square wave, see Witte, 2001.

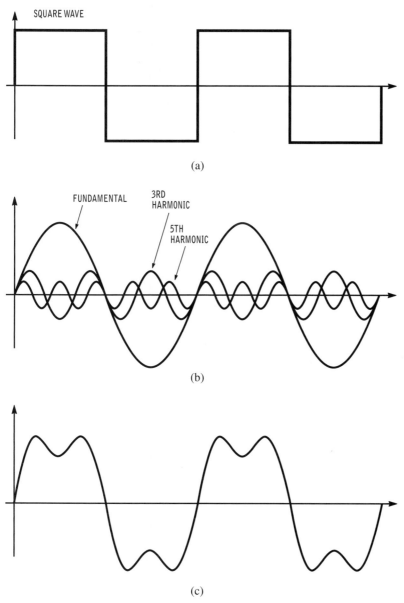

Figure 1.9 The square wave can be broken up into an infinite number of odd harmonics. The more harmonics that are included, the more the waveform approximates a square wave. a) The original square wave. b) The fundamental, third harmonic, and fifth harmonic. c) The fundamental plus third harmonic. c) The fundamental plus the third and fifth harmonics. d) The fundamental plus the third, fifth, and seventh harmonics. e) The fundamental plus the third, fifth, seventh, and ninth harmonics.

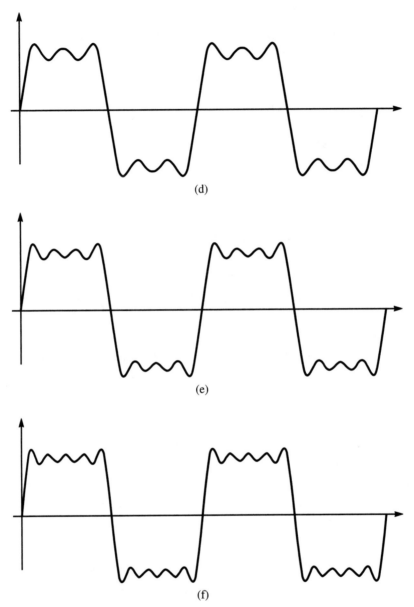

(d)

(e)

(f)

Figure 1.9   (Continued)

frequency domain representation or *spectrum* of the waveform. In the frequency domain representation, a vertical line (called a *spectral line*) indicates a particular frequency that is present (the fundamental or a harmonic). The height of each spectral line corresponds to the amplitude of that particular harmonic. A pure sine wave would be represented by one single spectral line. Figure 1.10 shows the frequency domain representation of a square wave. Notice that only the fundamental and odd harmonics are present and that each harmonic is smaller than the previous one. Understanding the spectral content of waveforms is important because the measurement instrument must be capable of operating at the frequencies of the harmonics (at least the ones that are to be included in the measurement).

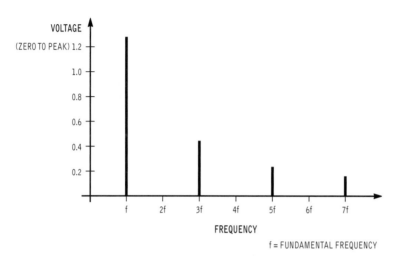

Figure 1.10   The frequency domain representation of a square wave, shown out to the seventh harmonic.

1.15   Pulse Train

The *pulse train*, or *repetitive pulse*, is a common signal in digital systems (Figure 1.11). It is similar to the square wave, but does not have both positive and negative values. Instead it has two possible values: V_{0-P} and 0 volts. The square wave spends 50% of the time at its positive voltage and 50% of the time at its negative voltage, corresponding to a 50% *duty cycle*. The pulse train's duty cycle may be any value between 0 and 100 percent and is defined by the following equation:

$$Duty\ Cycle \; = \; \frac{\tau}{T}$$

where τ is the length of time that the waveform is high and T is the period of the waveform.

The pulse train generates harmonic frequencies with amplitudes that are dependent on the duty cycle. A frequency domain plot of a typical pulse train (duty cycle of 25%) is shown in Figure 1.12. The envelope of the harmonics has a distinct humped shape that equals zero at integer multiples of $1/\tau$. Most of the waveform's energy is contained in the harmonics falling below this $1/\tau$ point. Therefore, it is often used as a rule of thumb for the bandwidth of the waveform.

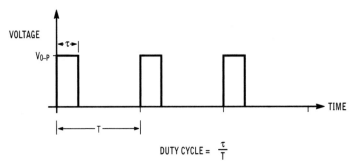

Figure 1.11 The pulse train is a common waveform in digital systems. The duty cycle describes the percent of the time that the waveform is at the higher voltage.

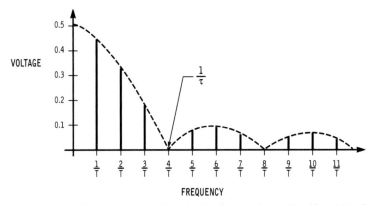

Figure 1.12 The frequency domain plot for a pulse train with a 25% duty cycle. The envelope of the harmonics falls to zero when the frequency equals 1/t.

Table 1.3 summarizes the harmonic content for each of the various waveforms with $V_{0-P} = 1$ volt. In all of the waveforms except the sine wave, there are an infinite number of harmonics but the amplitudes of the harmonics tend to decrease as the harmonic number increases. At some point, the higher harmonics can be ignored for practical systems because they are so small. As a measure of how wide each waveform is in the frequency domain, the last column lists the number of significant harmonics (ones that are at least 10% of the fundamental). The larger the number of significant harmonics, the wider the signal is in the frequency domain. The sine wave, of course, only has one significant harmonic (the fundamental can be considered the "first" harmonic). Notice that the amplitude value of the fundamental frequency component changes depending on the waveform, even though all of the waveforms have $V_{0-P} = 1$ volt.

Table 1.3 Table of harmonics for a variety of waveforms. All waveforms have a zero-to-peak value of 1. The number of significant harmonics column lists the highest harmonic whose amplitude is at least 10% of the fundamental.

| Waveform | Fund. | Harmonics | | | | | | Sig. Harm. (10%) | Equation |
		2nd	3rd	4th	5th	6th	7th		
Sine wave	1.000	0.000	0.000	0.000	0.000	0.000	0.000	1	
Square	1.273	0.000	0.424	0.000	0.255	0.000	0.182	9	$\dfrac{4}{n\pi}$ for odd n
Triangle	0.811	0.000	0.090	0.000	0.032	0.000	0.017	3	$\dfrac{8}{n^2\pi^2}$ for odd n
Pulse (50% duty cycle)	0.637	0.000	0.212	0.000	0.127	0.000	0.091	9	$\dfrac{2}{n\pi}\sin\left(\dfrac{n\pi}{2}\right)$
Pulse (25% duty cycle)	0.450	0.318	0.150	0.000	0.090	0.105	0.064	14	$\dfrac{2}{n\pi}\sin\left(\dfrac{n\pi}{4}\right)$
Pulse (10% duty cycle)	0.197	0.187	0.172	0.151	0.127	0.101	0.074	26	$\dfrac{2}{n\pi}\sin\left(\dfrac{n\pi}{10}\right)$

There are two important concepts to be obtained from the previous discussion of harmonics.

1. The more quickly a waveform transitions between its minimum and maximum values, the more significant harmonics it will have and the higher the frequency content of the waveform. For example, the square wave (which has abrupt voltage changes) has more significant harmonics than the triangle wave (which does not change nearly as quickly).
2. The narrower the width of a pulse, the more significant harmonics it will have and the higher the frequency content of the waveform.

Both of these statements should make intuitive sense if one considers that high frequency signals change voltage faster than lower frequency signals. The two conditions cited above both involve waveforms changing voltage in a more rapid manner. Therefore, it makes sense that waveforms that must change rapidly (those with short rise times or narrow pulses) will have more high frequencies present in the form of harmonics.

Example 1.5

What is the highest frequency that must be included in the measurement of a triangle wave that repeats every 50 microseconds (assuming that harmonics less than 10% of the fundamental can be ignored)?

The fundamental frequency, $f = 1/T = 1/50\,\mu\text{sec} = 20\,\text{kHz}$

From Table 1.3, the number of significant harmonics for a triangle wave (using the 10% criterion) is 3. The highest frequency that must be included is $3\,f = 3\,(20\,\text{kHz}) = 60\,\text{kHz}$.

1.16   Combined DC and AC

There are many cases where a waveform is not purely DC nor purely AC, but can be thought of as being a combination of the two. For example, in transistor circuits, an AC waveform is often superimposed on a DC bias voltage, as shown in Figure 1.13. The DC component (a straight horizontal line) and the AC component (a sine wave) combine to form the new waveform.

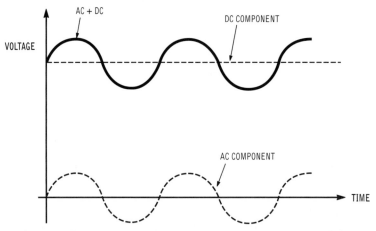

Figure 1.13 The waveform shown can be broken down into a DC component and an AC component. The DC component is just the average value of the original waveform.

The DC value of a waveform is also just the waveform's average value. In Figure 1.13, the waveform is above its DC value half of the time and below the DC value the other half, so on the average the voltage is just the DC value.

The pulse train shown in Figure 1.14 has a 50% duty cycle and is always positive in value. The average value of this waveform is, therefore, greater than zero. This waveform spends half of the time at V_{0-P} and the other half at 0, so the average or DC value =

$(V_{0-P} + 0)/2 = 1/2\ V_{0-P}$. The AC component left over when the DC component is removed is a square wave with half the zero-to-peak value of the original pulse train. In summary, the 50% duty cycle pulse train is equivalent to a square wave of half the zero-to-peak voltage plus a $1/2\ V_{0-P}$ DC component.

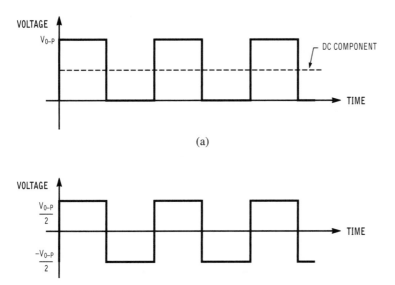

(a)

(b)

Figure 1.14 The 50% duty cycle pulse train is equivalent to a square wave plus a DC voltage. a) The pulse train with DC component shown. b) The square wave that results when the DC component is removed from the pulse train.

Example 1.6

A DC power supply has some residual AC riding on top of its DC component, as shown in Figure 1.15. Assuming that the AC component can be measured independently of the DC component, determine both the DC and AC values that would be measured (given the RMS value for the AC component).

The DC value is simply the average value of the waveform. Since the AC component is symmetrical, the average value can be calculated:

DC value = (10.6 + 10.2)/2 = 10.4 volts
If the DC is removed from the waveform, a triangle wave
with $V_{0-P} = 0.2$ volts is left.

For a triangle wave, $V_{RMS} = 0.577\ V_{0-P} = 0.115$ volts RMS

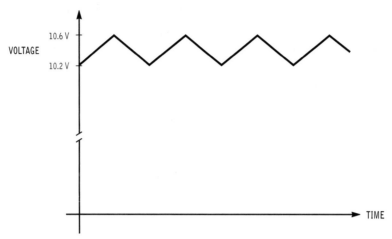

Figure 1.15 DC voltage with a residual triangle wave voltage superimposed on top of it (Example 1.6).

1.17 Modulated Signals

Sometimes sine waves are modulated by another waveform. For example, communication systems use this technique to superimpose low-frequency voice or data signals onto a high-frequency carrier that can be transmitted long distances. Modulation is performed by modifying some parameter of the original sine wave (called the *carrier*), depending on the value of the modulating waveform. In this way, information from the modulating waveform is transferred to the carrier. The most common forms of modulation used are amplitude, frequency, and phase modulation.

1.17.1 Amplitude Modulation

In *amplitude modulation* (AM), the amplitude of the carrier is determined by the modulating waveform. An amplitude modulated signal is represented by the equation

$$v(t) = A_c[1 + am(t)]\cos(2\pi f_c t)$$

where

A_c = the signal amplitude

a = modulation index ($0 \le a \le 1$)

$m(t)$ = normalized modulating signal (maximum value is 1)

f_c = carrier frequency

Figure 1.16a shows the original sine wave carrier with constant amplitude. Figure 1.16b is the modulating waveform. Figure 1.16c is the modulated carrier with the modulating waveform determining the amplitude of the carrier such that the modulating waveform

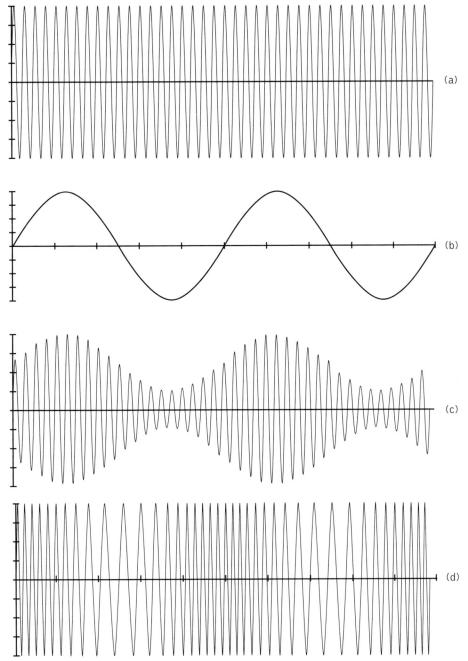

Figure 1.16 Amplitude and frequency modulation. a) The original carrier. b) The modulating signal. c) The resulting signal if amplitude modulation is used. d) The resulting signal if frequency modulation is used.

can be seen as the envelope of the modulated waveform. When the modulating waveform increases, the amplitude of the carrier increases and when the modulating waveform decreases, the amplitude of the carrier decreases.

1.17.2 Frequency Modulation

Frequency modulation (FM) also modulates a carrier, but the amplitude of the carrier remains constant while the frequency of the carrier changes. When the modulating waveform increases, the carrier frequency increases; when the modulating waveform decreases, the instantaneous carrier frequency decreases. A frequency-modulated signal is shown in Figure 1.16d.

Phase modulation is similar to frequency modulation. The carrier amplitude remains constant and the phase is changed according to the modulating waveform. Adjusting the phase of the carrier is somewhat like changing its frequency, so the effect on the carrier is similar. For many practical applications, phase and frequency modulation may be indistinguishable. FM and PM are generally considered to be variants of a type of modulation called *angle modulation*.

1.17.3 Modulated Signals in the Frequency Domain

Modulating a carrier, either with AM or FM, has an effect in the frequency domain (Figure 1.17). With no modulation, the carrier is just a pure sine wave and exists only at one frequency (Figure 1.17a). When the modulation is added, the carrier is accompanied by a band of other frequencies, called *sidebands*. The exact behavior of these sidebands depends on the level and type of modulation, but in general the sidebands spread out on one or both sides of the carrier (Figure 1.17b). Often they are relatively close to the carrier frequency, but in some cases, such as wideband FM, the sidebands spread out much farther. In all cases, modulation has the effect of spreading the carrier out in the frequency domain, occupying a wider bandwidth.

When a carrier is amplitude modulated by a single sine wave with frequency f_m, two modulation sidebands appear (offset by f_m) on both sides of the carrier (Figure 1.17c). This can be shown mathematically by inserting a sinusoid with frequency f_m into the equation for an amplitude-modulated signal:

$$v(t) = A_c[1 + a\cos(2\pi f_m t)]\cos(2\pi f_c t)$$

Using trigonometric identities, this equation can be rewritten as

$$v(t) = A_c\cos(2\pi f_c t) + \frac{aA_c}{2}\cos[2\pi(f_c + f_m)t] + \frac{aA_c}{2}\cos[2\pi(f_c - f_m)t]$$

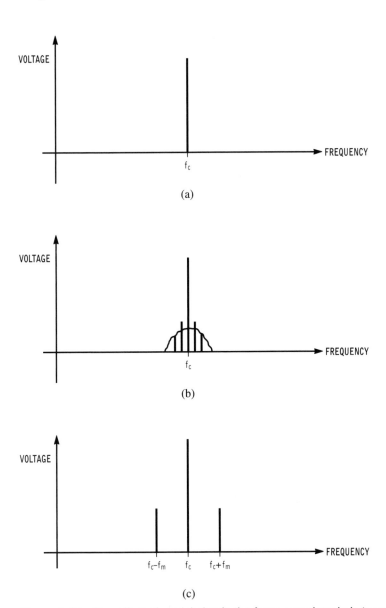

Figure 1.17 The effect of modulation in the frequency domain is to spread out the signal by creating sidebands. a) The unmodulated carrier is a single spectral line. b) Modulation on the carrier spreads the signal out in the frequency domain. c) An amplitude-modulated carrier (with single sine wave modulation) has a single pair of sidebands. d) A frequency-modulated carrier (with single sine wave modulation) may have many pairs of sidebands.

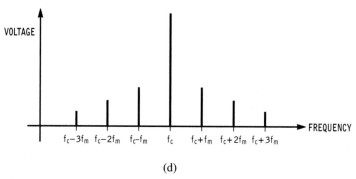

(d)

Figure 1.17 (Continued)

which gives us a frequency domain representation as consisting of the carrier frequency, f_c plus two sidebands at frequencies $f_c + f_m$ and $f_c - f_m$. The amplitudes of the two sidebands are the same and they depend on the modulation index, a.

Carriers that are frequency modulated may have a considerably larger number of side-bands.[6] With a modulating frequency of f_m, the sidebands appear at integer multiples of f_m away from the carrier frequency. In the general case, FM is a wideband form of modulation, in that the modulated carrier has sidebands that extend out farther than fm away from the carrier. If the modulation occurs at a low level, FM may be narrowband, with sidebands very similar to the AM case. For both modulation schemes, when the modulating signal is complex (voice waveforms, multiple sine waves, etc.), the sidebands are also more complex, occupying a continuum of frequencies adjacent to the carrier.

Communications systems represent intentional use of modulation but sometimes modulation is produced unintentionally due to circuit imperfections. In this case, the modulation is undesirable, or at least unnecessary. The effect in the frequency domain is still the same, however, so a sine wave with some residual modulation on it have sidebands on it in the frequency domain, causing the signal to spread out.

---

6.   The mathematical analysis of frequency modulation is complex and will not be covered in this book. For more information on FM, see Schwartz, 1993.

1.18 Decibels

The decibel (dB) is sometimes used to express electrical quantities in a convenient form. The definition of the decibel is based on the ratio of two power levels (log indicates the base ten logarithm):

$$dB = 10 \log\left[\frac{P_2}{P_1}\right]$$

Since $P = V^2/R$ (assuming RMS voltage),

$$dB = 10 \log\left[\frac{V_2^2/R_2}{V_1^2/R_1}\right]$$

And if $R_1 = R_2$ (the resistances involved are the same)

$$dB = 10 \log\left[\frac{V_2}{V_1}\right]^2$$

$$dB = 20 \log\left[\frac{V_2}{V_1}\right]$$

The voltages, V_1 and V_2, are normally RMS voltages. If the two waveforms are the same shape then the voltages can be expressed as zero-to-peak or peak-to-peak. Since many of our measurements are voltage measurements, calculating dB using voltages is the most convenient method. Strictly speaking, the voltage equation is valid only if the two resistances involved are the same. Misleading results can occur when $R_1 \neq R_2$ and the decibel equation is applied to both voltage and power. For example, the voltage gain calculation (in dB) of an amplifier often involves different resistances at the input and the output. However, it is common practice to apply the decibel equation to the voltage gain even though the two resistances are unequal.

To convert dB values back into a voltage or power ratio, a little reverse math is required:

$$\frac{P_2}{P_1} = 10^{(dB/10)}$$

$$\frac{V_2}{V_1} = 10^{(dB/20)}$$

The usefulness of decibels is sometimes questioned but they are widely used and must be understood by instrument users for that reason alone. In addition, they have at least two characteristics that make them very convenient:

1. Decibels compress widely varying electrical values onto a more manageable logarithmic scale. The range of powers extending from 100 watts down to 1 microwatt is a ratio of 100,000,000, but is expressed in dB as only 80 dB.
2. Gains and losses through circuits such as attenuators, amplifiers, and filters, when expressed in dB can be added together to produce the total gain or loss. To perform the equivalent operation without decibels requires multiplication.

Some cardinal decibel values are worth pointing out specifically:

- 0 dB corresponds to a ratio of 1 (for both voltage and power). A circuit that has 0-dB gain or 0-dB loss has an output equal to the input.
- 3 dB corresponds to a power ratio of 2. A power level that is changed by –3 dB is reduced to half the original power. A power level that is changed by +3 dB is doubled.
- 6 dB corresponds to a voltage ratio of 2. A voltage that is changed by –6 dB is reduced to half the original voltage. A voltage that is changed by +6 dB is doubled.
- 10 dB corresponds to a power ratio of 10. This is the only point where the dB value and the ratio value are the same (for power).
- 20 dB corresponds to a voltage ratio of 10. A voltage that is changed by +20 dB becomes 10 times larger. A voltage that is changed by –20 dB becomes 10 times smaller.

1.19   Absolute Decibel Values

Besides being useful for expressing ratios of powers or voltages, decibels can be used for specifying absolute voltages or powers. Either a power reference or voltage reference must be specified. For power calculations:

$$dB(absolute) = 10 \log\left(\frac{P}{P_{REF}}\right)$$

and for voltage calculations:

$$dB(absolute) = 20 \log\left(\frac{V}{V_{REF}}\right)$$

1.19.1 dBm

A convenient and often-used power reference for instrumentation use is the milliwatt (1 mW or 0.001 watt), which results in dBm:

$$dBm = 10 \log\left(\frac{P}{0.001}\right)$$

This expression is valid for any impedance or resistance value. If the impedance is known and specified, dBm can be computed using voltage. For a 50 Ω resistance, 1 mW of power corresponds to a voltage of 0.224 volts RMS:

$$P = \frac{V_{RMS}^2}{R}$$

$$V_{RMS} = \sqrt{P \cdot R} = \sqrt{0.001 \cdot 50} = 0.224 \text{ volts RMS}$$

Using this voltage as the reference value in the decibel equation results in:

$$dBm (50 \ \Omega) = 20 \log (V_{RMS} / 0.224)$$

Similarly, for 600 ohms and 75 ohms:

$$dBm (600 \ \Omega) = 20 \log (V_{RMS} / 0.775)$$

$$dBm (75 \ \Omega) = 20 \log (V_{RMS} / 0.274)$$

These equations are valid only for the specified impedance or resistance.

1.19.2 dBV

Another natural reference to use in measurements is 1 volt (RMS). This results in the following:

$$dBV = 20 \log (V/1)$$

or simply

$$dBV = 20 \log (V)$$

This equation is valid for any impedance level, since its reference is a voltage. Table 1.4 summarizes these equations and Figures 1.18 and 1.19 are plots of the basic decibel relationships. Table 1.5 shows voltage and power ratios along with their decibel values. The reader is encouraged to spend some time studying these figures to get a feel for how decibels work.

Table 1.4   Summary of equations relating to decibel calculations. Voltages (V) are RMS voltage and power (P) is in watts.

$$dB = 10 \log (P_2/P_1)$$

$$dB = 20 \log (V_2/V_1)$$

$$dBm = 10 \log (P/0.001)$$
$$dBV = 20 \log (V)$$

$$dBm\ (50\ \Omega) = 20 \log (V/0.224)$$
$$dBm\ (75\ \Omega) = 20 \log (V/0.274)$$
$$dBm\ (600\ \Omega) = 20 \log (V/0.775)$$
$$dBW = 10 \log (P)$$
$$dBf = 10 \log (P/1 \times 10^{-15})$$
$$dBuV = 20 \log (V/1 \times 10^{-6})$$
$$dBc = 10 \log (P/P_{carrier})$$
$$dBc = 20 \log (V/V_{carrier})$$

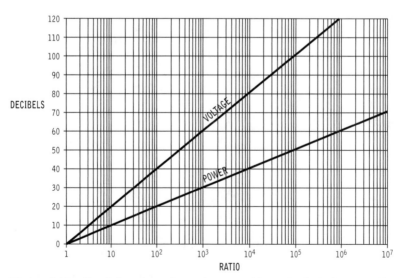

Figure 1.18   Graph for converting ratios to and from decibels for both voltage and power. Because of the logarithmic relationships, the plots are straight lines on log axes.

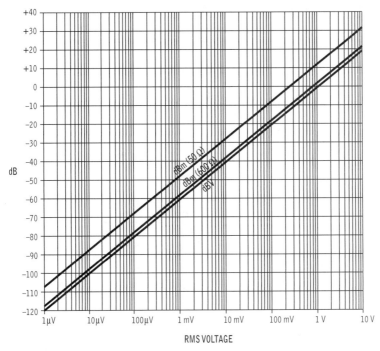

Figure 1.19 Graph of decibels versus voltage for dBV, dBm (50 Ω), and dBm (600 Ω).

Table 1.5 Table of decibel values for voltage ratios and power ratios.

Decibels	Power Ratio	Voltage Ratio
100	10000000000	100000
90	1000000000	31623
80	100000000	10000
70	10000000	3162
60	1000000	1000
50	100000	316.2
40	10000	100
30	1000	31.62
20	100	10
10	10	3.162
0	1	1.000
−10	0.1	0.3162

Table 1.5   Table of decibel values for voltage ratios and power ratios. (Continued)

Decibels	Power Ratio	Voltage Ratio
−20	0.01	0.1000
−30	0.001	0.03162
−40	0.0001	0.01000
−50	0.00001	0.003162
−60	0.000001	0.001000
−70	0.0000001	0.0003162
−80	0.00000001	0.0001000
−90	0.000000001	0.00003162
−100	0.0000000001	0.00001000
10	10.0000	3.1623
9	7.9433	2.8184
8	6.3096	2.5119
7	5.0119	2.2387
6	3.9811	1.9953
5	3.1623	1.7783
4	2.5119	1.5849
3	1.9953	1.4125
2	1.5849	1.2589
1	1.2589	1.1220
0.9	1.2303	1.1092
0.8	1.2023	1.0965
0.7	1.1749	1.0839
0.6	1.1482	1.0715
0.5	1.1220	1.0593
0.4	1.0965	1.0471
0.3	1.0715	1.0351
0.2	1.0471	1.0233
0.1	1.0233	1.0116

Table 1.5   Table of decibel values for voltage ratios and power ratios. (Continued)

Decibels	Power Ratio	Voltage Ratio
0	1.0000	1.0000
–0.1	0.9772	0.9886
–0.2	0.9550	0.9772
–0.3	0.9333	0.9661
–0.4	0.9120	0.9550
–0.5	0.8913	0.9441
–0.6	0.8710	0.9333
–0.7	0.8511	0.9226
–0.8	0.8318	0.9120
–0.9	0.8128	0.9016
–1	0.7943	0.8913
–2	0.6310	0.7943
–3	0.5012	0.7079
–4	0.3981	0.6310
–5	0.3162	0.5623
–6	0.2512	0.5012
–7	0.1995	0.4467
–8	0.1585	0.3981
–9	0.1259	0.3548
–10	0.1000	0.3162

Example 1.7

A 0.5-volt RMS voltage is across a 50 Ω resistor. Express this value in dBV and dBm. What would the values be if the resistor were 75 Ω?

50 Ω case:
    dBV = 20 log (0.5) = –6.02 dBV
    dBm (50 Ω) = 20 log (0.5/0.224) = +6.97 dBm
75 Ω case:
    dBV is based on voltage so that value is the same as the 50 Ω
      case (–6.02 dBV).

Example 1.7 (Continued)

dBm is referenced to 1 mW of power. We must either use the voltage equation for dBm (75 Ω) or compute the power and use the power equation. We will compute the power.

$$P = \frac{V_{RMS}^2}{R} = \frac{(0.5)^2}{75} = 3.33 \, \text{mW}$$

dBm = 10 log (P/0.001) = 10 log (0.00333/0.001)
dBm = 5.23 dBm

1.19.3 Other References

Other power reference values are 1 watt (dBW) and 1 femtowatt or 1×10^{-15} watt (dBf). A common voltage reference is the microvolt, resulting in dBμV. These values can be computed by the following equations:

$$dBW = 10 \log (P)$$

$$dBf = 10 \log (P/1 \times 10^{-15})$$

$$dB\mu V = 20 \log (V/1 \times 10^{-6})$$

Voltage or power measurements relative to a fixed signal or carrier are expressed as dBc.

$$dBc = 10 \log (P/P_{carrier})$$

or

$$dBc = 20 \log (V/V_{carrier})$$

The reader will undoubtedly encounter other variations. The number of possible permutations on the versatile decibel is limited only by the imagination of engineers worldwide.[7]

1.20 Measurement Error

A measuring instrument will always disturb the circuit being measured by removing energy from that circuit, no matter how small, and some error will always be introduced. Another way of saying this is that connecting an instrument to a circuit changes that circuit and changes the voltage or current that is being measured. This error can be minimized by careful attention to the loading effect, as discussed later.

---

7.   It is rumored that some electrical engineers balance their checkbooks using dB$.

1.20.1 Internal Error

Errors internal to the instrument also degrade the quality of the measurement. These errors fall into two main categories:

> **Accuracy**: the ability of the instrument to measure the true value to within some stated error specification.
>
> **Resolution:** the smallest change in value that an instrument can detect.

Suppose a voltage measuring instrument has an accuracy of ±1% of the measured voltage and a 3-digit resolution. If the measured voltage was 5 volts, then the accuracy of the instrument is 1% of 5 volts or ±0.05 volts. The instrument could read anywhere from 4.95 to 5.05 volts with a resolution of 3 digits. The meter cannot, for instance, read 5.001 volts, since that would require 4 digits of resolution.

If the instrument had 4-digit resolution (but 1% accuracy), then the reading could be anywhere from 4.950 to 5.050 volts (the same basic accuracy, but with more digits). Assume for the moment that both meters read exactly 5 volts. If the actual voltage changed to 5.001 volts, the 3-digit instrument would probably not register any change, but the 4-digit instrument would have enough resolution to show that the voltage had indeed changed. (Actually, the 3-digit meter reading might change, but could not display 5.001 volts because the next-highest possible reading is 5.01 volts.) The accuracy of the 4-digit instrument is not any better, but it has finer resolution and can detect smaller changes. Neither meter guarantees that a voltage of 5.001 volts can be measured any more accurately than 1%.

The example given is a digital instrument, but the same concepts apply to analog instruments. Resolution is not usually specified in digits in the analog case, but all instruments have some fundamental limitation to their measurement resolution. Typically, the limitation is the physical size of the analog meter and its markings. For instance, an analog voltmeter with full-scale equal to 10 volts will have difficulty detecting the difference between two voltages, which differ by a few millivolts.

Usually, an instrument provides more measurement resolution than measurement accuracy. This guarantees that the resolution will not limit the obtainable accuracy and allows detection of small changes in readings, even if those readings have some absolute error in them. The relative accuracy between small steps may be much better than the full-scale absolute accuracy.

1.21 The Loading Effect

In general, when two circuits (a source and a load) are connected together, the voltages and currents in those circuits both change. The source might be, for example, the output of an amplifier, transmitter, or signal generator. The corresponding load might be a speaker, antenna, or the input of a circuit. In the case where an electronic measurement is being made, the circuit under test is the source and the measuring instrument is the load.

Many source circuits can be represented by a simple circuit model called a *Thevenin equivalent circuit*. A Thevenin equivalent circuit is made up of a voltage source, V_S, and a series resistance, R_S (Figure 1.20). Thus, complex circuits can be simplified by replacing them with an equivalent circuit model. In the same way, many loads can be replaced conceptually with a circuit model consisting of a single resistance, R_L. (Resistive circuits will be considered initially in this discussion, but later expanded to include AC impedances.)

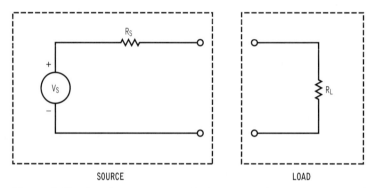

Figure 1.20 A voltage source with internal resistance and a resistive load (Thevenin equivalent circuit).

V_S is also known as the *open-circuit voltage*, since it is the voltage across the source circuit when no load is connected to it. This is easily proven by noting that no current can flow through R_S under open-circuit conditions. There is no voltage drop across R_S so the source voltage equals V_S.

1.22 The Voltage Divider

When the load is connected to the source, V_L, the voltage across the source and the load is no longer the open-circuit value, V_S. V_L is given by the voltage divider equation (named for the manner in which the total voltage in the circuit divides across R_S and R_L).

$$V_L = \frac{V_S R_L}{(R_S + R_L)}$$

Figure 1.21 shows a source and load connected in this manner. The resulting output voltage, V_L, as a function of the ratio R_L/R_S is plotted in Figure 1.22. If R_L is very small compared to R_S, then V_L is also very small. For large values of R_L (compared to R_S), V_L approaches V_S.

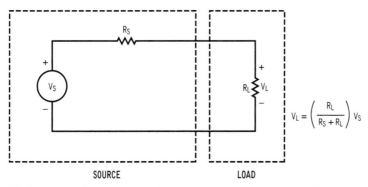

Figure 1.21 The source and load are connected together. The voltage across the load is given by the voltage divider equation. The loading effect causes this voltage to be less than the open circuit voltage of the source.

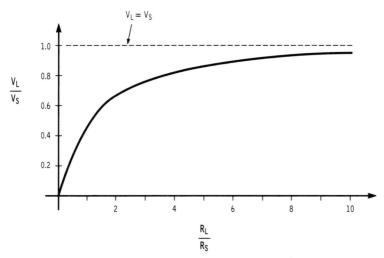

Figure 1.22 A plot of the output voltage due to the loading effect, as a function of the load resistance divided by the source resistance. The larger the load resistance, the better the voltage transfer.

1.23 Maximum Voltage Transfer

To get the maximum voltage out of a voltage source being loaded by some resistance, the ratio R_L/R_S should be as large as possible. From a design point of view, this can be approached from two directions: make R_S small or make R_L large. Ideally, we could make $R_S = 0$ and make $R_L = $ infinity, resulting in $V_L = V_S$. In practice, this cannot be obtained but can be approximated. Figure 1.22 shows that making R_L 10 times larger than R_S results in a voltage that is 91% of the maximum attainable (V_S).

When making measurements, the source is the circuit under test and the load is the measuring instrument. We may not have control over the value of R_S (if it is part of the circuit under test) and our only recourse is choosing R_L (the resistance in our measuring instrument). Ideally, we would want the resistance of the instrument to be infinite, causing no loading effect. In reality, we will settle for an instrument that has an equivalent load resistance that is much larger than the equivalent resistance of the circuit.

1.24 Maximum Power Transfer

Sometimes the power delivered to the load resistor is more important than the voltage. Since power depends on both voltage and current ($P = V \times I$), maximum voltage does not guarantee maximum power. The power delivered to R_L can be determined as follows:

$$P = \frac{V_S R_L}{(R_L + R_S)} \frac{V_S}{(R_L + R_S)}$$

$$P = \frac{V_S^2 R_L}{(R_L + R_S)^2}$$

This relationship is plotted for varying values of R_L/R_S in Figure 1.23. For small values of R_L (relative to R_S), the power delivered to R_L is small, because the voltage across R_L is small. For large values of R_L (relative to R_S), the power delivered is also small, because the current through R_L is small. The power transferred is maximum when $R_L/R_S = 1$, or equivalently, when R_L equals R_S.

$$R_L = R_S \text{ for maximum power}$$

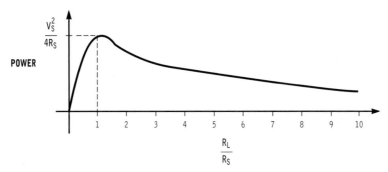

Figure 1.23 Plot of output power as a function of the load resistance divided by the source resistance. The output power is maximum when the two resistances are equal.

In many electronic systems, maximum power transfer is desirable. Such systems are designed with all source resistances and load resistances being equal to maximize the power transfer. Maximum power transfer is achieved while sacrificing maximum voltage transfer between the source and load.

1.25   Impedance

The previous discussion has assumed resistive circuits, but many circuit components are *reactive*, exhibiting a phase shift between their voltage and current. This type of voltage and current relationship is commonly represented through the use of *complex impedance*. The impedance of a device is defined as

$$Z = \frac{V_{0-P}\angle\theta_V}{I_{0-P}\angle\theta_I}$$

where

V_{0-P} is the AC voltage across the impedance
θ_V is the phase angle of the voltage
I_{0-P} is the AC current through the impedance
θ_I is the phase angle of the current.
Simplifying the equation,

$$Z = \frac{V_{0-P}}{I_{0-P}}\angle\theta_Z$$

where θ_Z is the phase angle of the impedance and is equal to $\theta_V - \theta_I$

The preceding equations show the complex impedance in magnitude and phase format. Alternatively, the impedance can be expressed in a rectangular format.[8]

$$Z = R + jX$$

where

R = the resistive component of the impedance
X = the reactive component of the impedance
j = the square root of -1

For the purposes of maximum power transfer, when the source impedance and load impedance are not resistive, maximum power transfer occurs when the load impedance has the same magnitude as the source impedance, but with opposite phase angle.[9] For instance, if the source impedance was 50 Ω with an angle of +45 degrees, the load impedance should

---

8.   See Appendix B for information on rectangular and polar format numbers.
9.   For a derivation of this, see Irwin, 2001.

be 50 Ω with an angle of –45 degrees to maximize the power transfer. Mathematically, this can be stated as

$$Z_L = Z_S{}^*$$

where * indicates the complex conjugate.

A special case occurs when the phase angle of the source impedance is zero. In that case, the load impedance should have the same magnitude as the source impedance and an angle of zero (load impedance equals source impedance). Note that this is the same as the purely resistive case.

1.26   Instrument Inputs

The characteristics of instrument input impedances vary quite a bit with each individual model, but in general they can be put into two categories: *high impedance* and *system impedance.*

1.26.1   High-Impedance Inputs

High impedance inputs are designed to maximize the voltage transfer from the circuit under test to the measuring instrument by minimizing the loading effect. As outlined previously, this can be done by making the input impedance of the instrument much larger than the impedance of the circuit. Typical values for instrument input impedance are between 10 kΩ and 1 MΩ. For instruments used at high frequencies, the capacitance across the input becomes important and therefore is usually specified by the manufacturer.

1.26.2   System-Impedance Inputs

Many electronic systems have a particular system impedance such as 50 Ω (Figure 1.24). If all inputs, outputs, cables, and loads in the system have the same resistive impedance, then, according to the previous discussion, maximum power is always transferred. At high frequencies (greater than about 30 MHz), stray capacitance and transmission line effects make this the only type of system that is practical. The system impedance is often called the *characteristic impedance* and is represented by the symbol Z_0.

At audio frequencies, a constant system impedance is not mandatory but is often used. It is sufficient for many applications to make all source circuits have a low impedance (less than 100 Ω) and all load circuits have a high impedance (greater than 1 kΩ). This results in near-maximum voltage transfer (power transfer is less important here). Some audio systems do maintain, or at least specify for test purposes, a system impedance, which is usually 600 Ω. This same impedance appears in telephone applications as well.

For radio and microwave frequency work, 50 Ω is by far the most common impedance. This impedance can be easily maintained despite stray capacitance and 50 Ω cable is easily realizable. Such things as amateur and commercial radio transmitters, transmitting

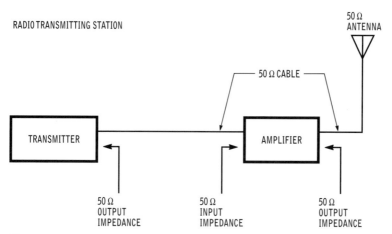

Figure 1.24 An electronic system that uses a common system impedance for maximum power transfer throughout.

antennas, communications filters, and radio frequency test equipment all commonly have 50 Ω input and output impedances. Runner up to 50 Ω in the radio frequency world is 75 Ω. This impedance is also used extensively at these frequencies, particularly in video-related applications such as cable TV. Other system impedances are used as special needs arise and will be encountered when making electronic measurements.

When measuring this type of system, most of the accessible points in the system expect to be loaded by the system impedance (Z_0). Because of this, many instruments are supplied with standard input impedance values. Thus, the instrument can be connected into the system and act as a Z_0 load during the measurement.

1.26.3 Connectors

Instruments use input and output connectors that are consistent with the signal or waveform being generated or measured. Usually, it is the frequency range of the instrument that determines the type of connector required.

For DC and low-frequency AC instruments, the *banana plug* and *banana jack* are often used, usually configured in pairs as a *dual-banana plug* or *jack*. The banana jack may be the "binding post" style that allows a bare wire to be connected to the jack without the need for a plug. For applications that could have high voltage present on the banana jack, such as with the input to a multimeter, the banana jack and plug are insulated to prevent hazardous voltages being exposed to the user.

The *BNC connector* is the most common connector used in electronic test equipment.[10] The BNC connector is compatible with coaxial cables and is easy to connect and disconnect due to its quarter-turn locking mechanism. The connector is available in 50 Ω and 75 Ω versions, consistent with the most common system impedances. These connectors can be used from DC to several GHz, depending on the specific application.

The *Type N connector* is considerably larger than the BNC connector and has a threaded locking mechanism.[11] The N connector is compatible with coaxial cables, is available in 50 Ω and 75 Ω versions, and is used for frequencies from DC to over 10 GHz. The threaded locking mechanism provides for a more repeatable connection than the BNC connector such that the N connector is sometimes used for precision measurements regardless of frequency.

1.27   Bandwidth

Instruments that measure AC waveforms generally have some maximum frequency above which the measurement accuracy is degraded. This frequency is the *bandwidth* of the instrument and is usually defined as the frequency at which the instrument's response has decreased by 3 dB. (Other values such as 1 dB and 6 dB are also sometimes used.) A typical response is shown in Figure 1.25. Note that the response does not instantly stop at the 3-dB bandwidth. It begins decreasing at frequencies less than the bandwidth and the response may still be usable for frequencies outside the bandwidth. Some instruments have their bandwidth implied in their accuracy specification. Rather than give a specific 3-dB bandwidth, the accuracy specification will be given as valid only over a specified frequency range.

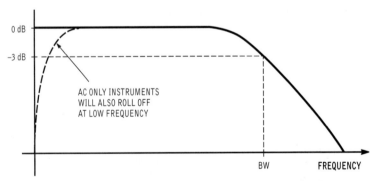

Figure 1.25   The frequency response of a typical measuring instrument rolls off at high frequencies. Some instruments that do not measure DC also roll off at low frequencies.

---

10.   BNC stands for Bayonet Neill Concelman, named for the inventors of the connector, Paul Neill and Carl Concelman.
11.   Type N stands for Neill, named for its inventor, Paul Neill.

In general, to measure an AC waveform accurately, the instrument must have a bandwidth that exceeds the frequency content of the waveform. For a sine wave, the instrument bandwidth must be at least as large as the sine wave frequency. For sine wave frequencies equal to the 3-dB bandwidth, the measured value will, of course, be decreased by 3 dB, so an even wider bandwidth is desirable. The amount of margin necessary will vary depending on how quickly the response rolls off near the 3-dB point.

For sine wave measurements, the frequency response near the 3-dB bandwidth can be adjusted for by recording the response at these frequencies using a known input signal. Assuming the response is repeatable, the measurement error due to limited bandwidth can be used to adjust the measured value to obtain the actual value.

Instruments that measure only AC voltages (and not DC) have a response that rolls off at low frequencies as well as high frequencies. When using instruments of this type (including many AC voltmeters), both the high-frequency and the low-frequency bandwidth limitations must be considered.

For waveforms other than sine waves, the harmonics must be considered. If the harmonics are outside the instrument bandwidth, their effect on the waveform will not be measured. This could be desirable if the harmonics outside some frequency range needed to be ignored. In general, their effect is usually desirable in the measurement. Waveforms with an infinite number of harmonics would require an instrument with infinite bandwidth to measure them. In reality, the higher harmonics have very little energy and can be ignored with proper regard to how much error this produces in the measurement.

Example 1.8

An electronic instrument is used to measure the voltage of a 2-kHz sine wave. What instrument bandwidth is required? What bandwidth would be required to measure a square wave having the same frequency (assume that any harmonic greater than 10% of the fundamental is to be included)?

Since a sine wave has only the fundamental frequency, the bandwidth of the instrument must be at least the frequency of the waveform. BW = 2 kHz. (In reality, we may want to choose a somewhat higher value since the instrument response is diminished by 3 dB at its bandwidth.)

From Table 1.3, the highest significant harmonic of a square wave (greater than 10% of the fundamental) is the ninth harmonic. Therefore, the bandwidth must be at least (and probably larger than):
$$BW = 9 \times 2 \text{ kHz} = 18 \text{ kHz}$$

1.28   Rise Time

Ideally, waveforms such as square waves and pulses change voltage level instantaneously. In reality, waveforms take some time to make an abrupt change, depending on the
bandwidth of the system and other circuit parameters. The amount of time it takes for a
waveform to transition from one voltage to another is called the *rise time*. (The rise time is
normally measured at the 10% and 90% levels of the transition.) The bandwidth of a measuring instrument will limit the measured rise time of a pulse or square wave. For a typical
instrument, the relationship between rise time and bandwidth is given by:

$$t_{RISE} = \frac{0.35}{BW}$$

BW = 3-dB bandwidth (in Hertz)

The validity of this relationship depends on the exact shape of the frequency response
of the instrument (how fast it rolls off for frequencies above its bandwidth). It is exact for
instruments with a single-pole roll-off[12] and is a good approximation for many instruments.
The important point here is that the bandwidth, which is a frequency domain concept, limits
the measurement of the rise time, which is a time domain concept. The two characteristics
of the instrument (time domain and frequency domain) are intertwined. Fast changes (small
rise times) correspond to high-frequency content in a waveform, so if the high-frequency
content is limited by the bandwidth of a system, then the rise time will be larger.

The instrument should have a rise time significantly smaller than the rise time being
measured. A rise time measurement using an instrument with a rise time two times smaller
than the one being measured will result in an error of about 10%. This error drops to 1%
when the instrument rise time is 7 times shorter than the measured rise time.

1.29   Bandwidth Limitation on Square Wave

As an example of how limited bandwidth can affect a waveform in the time domain,
consider the square wave shown in Figure 1.26. The waveform is passed through a low-pass
filter, which has a frequency characteristic similar to Figure 1.25. If the bandwidth of the filter is very wide (compared to the fundamental frequency of the waveform), the square wave
appears at the output undistorted. If the bandwidth is reduced, some of the harmonics are
removed from the waveform. The output still looks like a square wave, but it has some
imperfections that would cause a measurement error. For very limited bandwidth, the
square wave barely appears at all, and is very rounded due to the lack of high-frequency
harmonics.

---

12.   See Appendix B for a discussion of this type of frequency response and the derivation of the rela-
      tionship between bandwidth and rise time.

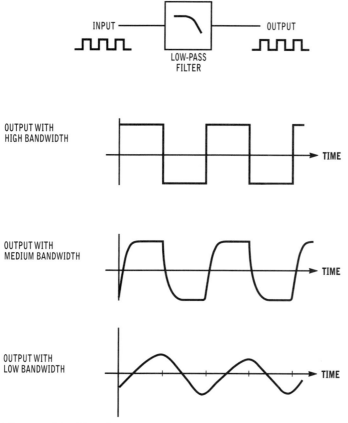

INPUT ——— [LOW-PASS FILTER] ——— OUTPUT

LOW-PASS
FILTER

OUTPUT WITH
HIGH BANDWIDTH

TIME

OUTPUT WITH
MEDIUM BANDWIDTH

TIME

OUTPUT WITH
LOW BANDWIDTH

TIME

Figure 1.26   The effect of bandwidth on a square wave. With a very wide bandwidth, the square wave is undistorted; with a low bandwidth, the square wave is distorted.

Example 1.9

A pulse with zero rise time is passed through a low-pass Filter with a 3-dB bandwidth of 25 kHz. If the filter has a single-pole roll-off, what will the rise time be at the output?

Even though the rise time at the input is zero, due to the 25 kHz bandwidth limitation the rise time at the output will be

$$t_{RISE} = 0.35/ BW = 0.35/ 25 \text{ kHz} = 14 \text{ } \mu\text{sec}$$

> **Example 1.10**
>
> **A square wave with a rise time of 1 usec is to be measured by an instrument. What 3-dB bandwidth is required in the measuring instrument to measure the rise time with 1% error?**
>
> For a 1% error in rise time, the measuring instrument must have a rise time that is 7 times less than the waveform's rise time.
>
> For the instrument:
>
> $$t_{RISE} = 1 \ \mu sec / 7 = 0.14 \ \mu sec$$
>
> $$BW = 0.35 / t_{RISE} = 0.35 / 0.14 \ \mu sec = 2.5 \ MHz$$

1.30 Digital Signals

With the widespread adoption of digital logic circuitry, digital signals have become very common in electronic systems. The basic advantage and simplicity of a digital signal is that it has only two valid states: HIGH and LOW (Figure 1.27). The HIGH state is defined as any voltage greater than or equal to V_H and the LOW state is defined as any voltage less than or equal to V_L. Any voltage between V_H and V_L is undefined. V_H and V_L are called the *logic thresholds*.

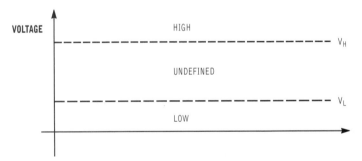

Figure 1.27   A digital signal must be one of two valid states, HIGH or LOW.

A typical, well-behaved digital signal (a pulse train) is shown in Figure 1.28a. Whenever the voltage is V_{0-P}, the digital signal is interpreted as HIGH. Whenever the voltage is zero, the digital signal is LOW. The digital signal in Figure 1.28b is not so well behaved. The signal starts out LOW, then enters the undefined region (greater than V_L but smaller than V_H), then becomes HIGH, then LOW and HIGH once more. Notice that in the last LOW state, the voltage is not zero but increases a small amount. It does stay less than V_L, so it is still a LOW signal. Similarly, during the last HIGH state the voltage steps down a small amount, but stays above the high threshold.

Although the region between the two logic thresholds is undefined, it does not follow that it should never occur. First of all, to get from LOW to HIGH, the waveform must pass through the undefined area. Also, there are times when one digital gate goes into an "off" or high impedance state while another gate drives the voltage HIGH or LOW. During this transition, the voltage could stay in the undefined area for a period of time. It is imperative that the digital signal settle to a valid logic level before the digital circuit following it uses the information, otherwise a logic error may occur.

Digital signals are mostly used to represent the binary numbers 0 and 1. If *positive logic* is being used, then a HIGH state corresponds to a logical 1 and a LOW state corresponds to a logical 0. Although positive logic may appear to be the most obvious convention, *negative logic* is also sometimes used. With negative logic, the HIGH state represents a logical 0 and the LOW state represents a logical 1. Thus, the relationship between the voltage of the digital signal and the binary number that it represents varies depending on the logic convention used. Fortunately, the concepts associated with digital signals remain the same but the instrument user may be left with a digital bookkeeping problem.

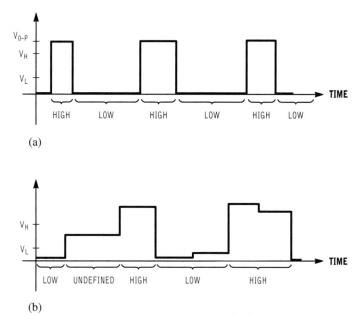

Figure 1.28 Two digital signals. a) A well-behaved digital signal that is always a valid logic level. b) A digital signal that is undefined at one point.

1.31 Logic Families

There are several different technologies used to implement digital logic circuits, with different speed and logic threshold characteristics. For any particular logic family, we need to know the correct values for the logic thresholds (V_H, V_L) so that the digital signal can be measured or interpreted correctly. Table 1.6 lists logic thresholds for the most popular logic families.

Table 1.6   Logic Levels for Standard Digital Logic Families

Logic Family	Supply Voltage V_{CC}	Input Threshold (Low), V_{IL}	Input Threshold (High), V_{IH}	Output Threshold (Low), V_{OL}	Output Threshold (High), V_{OH}
CMOS	5 V	1.5 V	3.5 V	0.5 V	4.4 V
TTL	5 V	0.8 V	2.0 V	0.4 V	2.4 V
LVTTL	3.3 V	0.8 V	2.0 V	0.4 V	2.4 V
LVCMOS	3.3 V	0.8 V	2.0 V	0.2 V	3.1 V
LVCMOS	2.5 V	0.7 V	1.7 V	0.4 V	2.0 V

Derived from "Benefits and Issues on Migration of 5-V and 3.3-V Logic to Lower-Voltage Supplies," Texas Instruments, September 1999.

The logic thresholds are different depending on whether you are concerned with the input of a gate or the output of a gate. The V_L for the output of a gate is lower than the V_L for the input to a gate. Similarly, the V_H for the output is higher than the V_H for the input. This slight amount of intentional design margin guarantees that the output of a digital circuit will drive the input of the next circuit well past its required logic threshold. The difference between the input threshold and the output threshold is called the *noise margin*, since the excess voltage is designed to overcome any electrical noise in the system. In general digital testing, the measurement instrumentation should usually be considered a digital input and the logic levels associated with the input should be used.

Standards and prevailing practice concerning logic signal levels are changing over time. The drive for more logic gates on an integrated circuit and higher-speed operation of those gates drives the technology toward smaller device size. As the number of devices on an integrated circuit increases, power consumption tends to also increase. To counteract both of these trends, integrated circuit manufacturers are specifying reduced power supply voltages on digital circuits. Lower supply voltages mean lower voltage stress on the devices, allowing them to be made smaller. A lower supply voltage also significantly decreases the power consumed by a given circuit.

The dominant logic levels in modern digital circuits were driven by the popularity of TTL (transistor–transistor logic) technology, implemented with a 5-volt power supply. In recent years, CMOS (Complementary Metal Oxide Semiconductor) has become the dominant digital technology. However, TTL logic levels are still commonly used. Logic gates may be described as "TTL-compatible," while actually implemented using CMOS technology. Historically, the most common power supply voltage for logic circuits is 5 volts, which was used on a series of popular TTL logic families. When CMOS technology gained in usage, the 5-volt power supply standard was adopted. More recent logic families use lower supply voltages, including 3.3-volt and 2.5-volt supplies. Backward compatibility with the TTL input high and low of 2.0 volts and 0.8 volts is possible with the 3.3-volt supply but not with the 2.5-volt supply.

These industry trends are creating confusion and change in digital logic levels. While there is a desire to maintain "TTL compatibility," the drive to lower supply voltages and higher speed is forcing changes in logic levels. Table 1.6 can be used as a guideline but users of test equipment may need to consult the data sheet for a particular part.

1.32   References

"2.5 V± 0.2 V (Normal Range), and 1.8 V to 2.7 V (Wide Range) Power Supply Voltage and Interface Standard for Nonterminated Digital Integrated Circuit," EIA/JEDEC Standard JESD8-5, Electronics Industry Alliance, October 1995.

"Interface standard for Nominal 3 V/3.3 V Supply Digital Integrated Circuits," EIA/JEDEC Standard JESD8-B, Electronics Industry Alliance, September 1999.

Irwin, J. David. *Basic Engineering Circuit Analysis*, 7th ed., New York: Wiley, 2001.

Oliver, Bernard M., and John M. Cage. *Electronic Measurements and Instrumentation*. New York: McGraw-Hill, 1971.

Schwartz, Mischa. *Information, Transmission, Modulation, and Noise*, 4th ed., New York: McGraw-Hill, 1990.

Schwarz, Steven E., and William G. Oldham. *Electrical Engineering: An Introduction*, 2nd ed., New York: Oxford University Press, 1993.

Wedlock, Bruce D., and James K. Roberge. *Electronic Components and Measurements*, Englewood Cliffs, NJ: Prentice Hall, 1969.

Witte, Robert A. *Spectrum and Network Measurements*, Norcross, GA: Noble, 2001.

CHAPTER 2

Voltmeters, Ammeters, and Ohmmeters

Meters are the simplest, easiest to use instruments for measuring voltage, current, and resistance. A *voltmeter* measures voltage; an *ammeter* measures current, and an *ohmmeter* measures resistance. These three meter functions are often combined into one instrument, known as a *multimeter.* Convenience features such as autoranging (automatic selection of measurement range) combine with very high input impedance and digital display to provide uncomplicated meter measurements. For many applications, meter measurements have become as simple as selecting the meter function and connecting to the circuit under test. However, as the capability of a meter is stretched, more attention must be paid to the details of the measurement.

2.1   Meters

Meters can be classified not only by the parameter that they measure, but also by the type of display or readout used. Meter displays fall into two categories: analog and digital.

2.1.1   Analog Meters

Analog meters generally use some sort of electromechanical mechanism to cause a small arm to move, depending on the voltage or current applied to the meter (Figure 2.1). A graduated measurement scale is imprinted behind the mechanism such that the moving arm points to the value of the meter reading. The actual mechanism involved is not critical; the important point is that analog meters provide a continuously varying readout, without any discrete jumps in meter reading. The mechanism does not have to be mechanical. It can be implemented using other technologies as long as the result is a continuous "analog" type of display.

Analog meters require care in observing meter readings. Different people watching the same measurement may disagree slightly on the correct interpretation of the meter reading. The resolution of the meter (the ability to measure small changes) depends on the physical layout of the meter, but generally is no smaller than a few percent of the full-scale meter reading. The biggest advantage that analog meters have when compared to digital meters is the ability to track changes in meter readings. Sometimes the ability to spot a trend in a meter reading (whether it is increasing or decreasing) is just as important as the absolute reading. For example, if a circuit is to be adjusted for maximum output voltage, an analog meter gives the user visual feedback for making the adjustment, and the output can be peaked very quickly.

Figure 2.1   An analog meter has a continuous scale. (Photo courtesy of B&K-Precision.)

2.1.2   Digital Meters

Digital meters, on the other hand, do not provide continuously variable meter readings. The meter reading is converted into decimal digits and is displayed as a number (Figure 2.2). This results in a reading that is very easy to interpret. Several different people viewing the same meter will record the same reading (except for, perhaps, cases where the number is changing due to electrical noise or measurement drift). Digital meters are usually specified according to the number of digits in the display. A 3-digit voltmeter has three digits in its display. A $3\frac{1}{2}$-digit voltmeter has three normal decimal digits plus a leading digit, which has a limited range. Typically, the leading digit might only have allowable values of 0, 1, 2, or 3. Some meters are more restrictive, with a leading digit that can only be one or zero. The range of the display is specified by the manufacturer, often as a maximum "count." For instance, a 3,200-count display on a 3-volt range can display voltages up to 3.200 volts.

Adjusting a circuit for maximum voltage is much more difficult with a digital meter than with an analog meter. The user must read the number, decide whether the new reading is larger or smaller than the previous reading, and then tweak the adjustment in the proper direction. For meters with three or four digits, this can be difficult. Some digital meters have included a simple analog-like bar graph display in addition to the high-resolution digital display to provide the best of both technologies. The digital readout can be used for precise absolute measurements while the bar graph display can be used to adjust for maximum or minimum value.

It is important not to assume that the presence of many display digits automatically means very high accuracy. Often, a meter will provide more resolution than its accuracy specification supports. This ensures that the resolution of the meter does not limit its accuracy, but may mislead some users. The amount of accuracy versus resolution may vary with the different operating modes of the instrument, so careful examination of the specification sheet is in order.

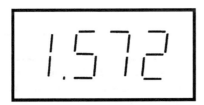

Figure 2.2   A digital meter displays the meter reading as a decimal number.

2.2   DC Voltmeters

The *ideal voltmeter* (Figure 2.3) senses the voltage across its terminals without drawing any current because it has an infinite internal resistance. Therefore, it does not load the circuit under test (and, therefore, it is impossible to implement an ideal voltmeter). In other words, the ideal voltmeter looks like an open circuit to the circuit under test. This is exactly what we want: a meter that can be connected to a circuit without disturbing it in any way. This circuit model is valid for both analog and digital meters.

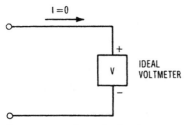

Figure 2.3   The ideal voltmeter senses the voltage across its terminals without drawing any current.

A more realistic model of a *real voltmeter* (Figure 2.4) includes the internal resistance of the meter. If the internal resistance of the meter is high enough compared with the resistance of the circuit under test, then there will be only minimal loading and the meter will approximate the ideal case. (See Chapter 1 for a discussion of the loading effect.)

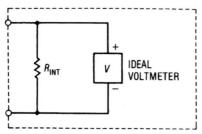

Figure 2.4   The real voltmeter has a finite internal resistance, which draws some current from the circuit under test.

Example 2.1

A voltmeter with an internal resistance of 100 kΩ is used to measure a 10-volt source having an internal resistance of 2 kΩ (Figure 2.5). What is the error in the measurement (due to loading)?

If the voltmeter were ideal and did not load the circuit (Figure 2.6a), the measured voltage would simply be the value of the voltage source, 10 volts. But with a real voltmeter (Figure 2.6b), the loading effect must be taken into account. Since the circuit is a voltage divider:

$$V_M = 10 \times 100 \text{ k}/(2 \text{ k} + 100 \text{ k}) = 9.804 \text{ volts (real voltmeter)}$$

$$\text{Error} = \text{real} - \text{ideal} = 9.804 - 10 = -0.196 \text{ volts}$$

The error is 2% of the ideal value.

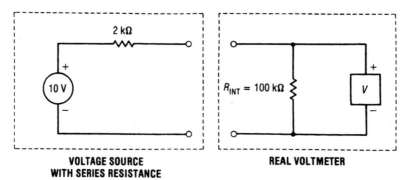

**VOLTAGE SOURCE
WITH SERIES RESISTANCE**

REAL VOLTMETER

Figure 2.5   A real voltmeter is used to measure the voltage source with an internal resistance.

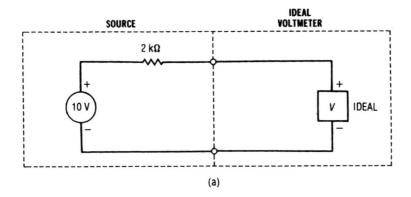

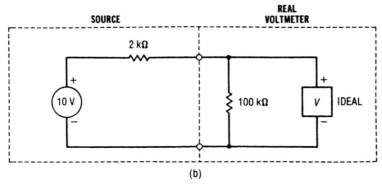

Figure 2.6 Measuring with an ideal voltmeter and a real voltmeter. a) With an ideal meter, no current is drawn and no voltage drop appears across the resistor. The meter reads 10 volts. b) With a real meter, some current is drawn and the loading effect reduces the measured voltage.

2.3 AC Voltmeters

The AC voltmeter is very similar to the DC voltmeter except that it measures the voltage of AC waveforms. (In some cases, it measures the combined DC and AC portions of a waveform.) The same ideal voltmeter and real voltmeter circuit models shown in Figure 2.3 and Figure 2.4 apply to the AC voltmeter. Again, the ideal voltmeter does not load the circuit under test, due to its infinite internal resistance. The loading effect of the real AC voltmeter is modeled by the resistance in Figure 2.4. A capacitor may be added in parallel with the resistance in the circuit model of the real meter to account for the input capacitance of the AC meter. (This is not a concern for DC meters since a capacitor acts like an open circuit for DC and does not load the circuit.)

2.3.1   Average-Responding Meter

Unless otherwise specified, AC voltmeters are usually calibrated to read out in RMS volts. Not all meters, however, actually measure the RMS value of the signal. Many low-cost meters use a circuit that responds to the average value of the waveform.[1] The meter readout is still calibrated to read RMS volts as long as the waveform is a sine wave. This is possible because the ratio of the average value and RMS value for a particular waveform can be determined. This is a perfectly valid technique as long as it is understood that the meter readout is only correct for sine waves. If a square wave were measured using an average responding meter, the meter reading would be invalid. Some average-responding meters respond to the average value of the *half-wave* rectified waveform (instead of the full-wave rectified waveform). These meters will also be in error when used on anything but sine waves.

Some manufacturers supply correction factors with their average-responding meters such that other waveforms can be measured. This obviously requires that the shape of the waveform be known, since each waveform will have a different correction factor. If waveforms other than sine waves must be measured using an average responding meter and the manufacturer has not supplied the correction factors, they may be determined experimentally by connecting the appropriate waveform with a known RMS voltage. This is a reliable technique as long as the shape and frequency of the waveform are not varied.

2.3.2   Peak-Reading Meters

Some AC voltmeters use a peak-detecting circuit that responds to the peak-to-peak value of the waveform. The meter display may be in terms of peak-to-peak volts or RMS volts. If the meter reads in peak-to-peak volts, the measured value should be accurate regardless of the shape of the waveform. However, if the meter reads RMS volts, then the reading is valid only for a sine wave. In a manner very similar to the average responding meter, the peak reading meter internally scales its measured voltage to obtain an RMS reading. This is done for only one waveform: the sine wave. Therefore, the peak-to-peak reading will be accurate independent of waveform, but the RMS reading is valid only for a sine wave.

2.3.3   True-RMS Meters

Meters that respond to the actual RMS value of the waveform are called *True-RMS meters.*[2] This simplifies the measurement, since the meter reads the correct RMS voltage regardless of the type of waveform. The root-mean-square function described in Chapter 1 is not simple to implement. Historically, these meters have been rather expensive to build,

---

1.   Here, average value refers to the full-wave rectified average value discussed in Chapter 1.
2.   The term "True RMS" is used to distinguish this type of meter from average-responding meters, which are calibrated to read RMS voltage.

especially if a wide bandwidth was required. As integrated circuit technology has advanced in recent years, the price of True RMS has decreased dramatically and this feature now appears in low-cost meters.

Even True RMS meters have some limitation on the shape of the waveform that can be measured. This limitation is usually specified as a maximum crest factor (peak-to-RMS ratio) that can be tolerated, perhaps with some specified accuracy. As a waveform's crest factor increases, a meter has to handle a larger peak voltage while measuring a relatively smaller RMS value.

2.3.4   Bandwidth

When making AC measurements, the bandwidth of the meter must be considered. Sometimes the bandwidth of the meter is specified directly, but more often the manufacturer provides an accuracy specification that depends on the frequency. Either way, the useful frequency range of the meter is defined.

For sine waves, the bandwidth must be at least as high as the frequency of the waveform. For other waveforms that contain harmonics, the harmonics must be included (see the discussion of bandwidth in Chapter 1).

2.3.5   AC and DC Coupling

For waveforms containing both DC and AC components, the AC voltage may be thought of as riding on top of the DC level. If an AC voltmeter is used to measure this type of voltage, the resulting meter reading will greatly depend on the design of the meter. Some manufacturers have included a coupling capacitor on the input of their voltmeters so that the meter is *AC coupled*. This AC coupling capacitor will block any DC voltage that is present, but lets the AC portion of the waveform pass through to the meter. So if a meter is AC coupled, only the AC portion of the voltage is measured. (This concept of AC and DC coupling also applies to other instruments, particularly oscilloscopes.)

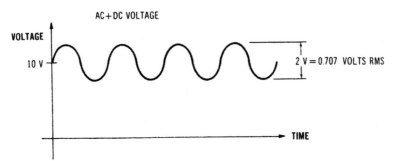

Figure 2.7   The measured value of a waveform containing both AC and DC will depend on the design of the meter. In this example, an AC-coupled meter will read 0.707 volts RMS and a DC-coupled meter will read 10.025 volts RMS.

Some instruments do not have a coupling capacitor and respond to both the DC and the AC voltages present. This is referred to as *DC coupling*. It is important to understand how a particular meter behaves if mixed DC and AC voltages are to be measured. Suppose the waveform in Figure 2.7 is to be measured by both types of meter. If the meter is AC coupled, then only the 2-volt peak-to-peak sine wave will be measured, resulting in V_{RMS} = 0.707 volts. On the other hand, if the meter is DC coupled, then both the AC and DC will be measured, resulting in a much higher reading. One might expect that the RMS reading in this case is just the sum of the DC and AC (RMS) voltages. This is not the case, since the RMS function of the voltmeter does not simply add the two voltages together. Assuming that the meter is a True RMS meter, it will read

$$V_{RMS} = \sqrt{V_{DC}^2 + V_{AC}^2} = \sqrt{10^2 + 0.707^2} = 10.025 \text{ volts}$$

If a meter is DC coupled, adding an external capacitor will make it AC coupled, as shown in Figure 2.8. The capacitor will cause the meter response to roll off at low frequencies. The 3-dB point will be given by:

$$f_{3dB} = \frac{1}{2\pi R_{INT} C}$$

where R_{INT} is the internal resistance of the voltmeter. The value of the capacitor should be chosen such that the 3-dB frequency is about 10 times smaller than the frequency of the waveform being measured. This ensures that the frequency being measured is well within the bandwidth of the meter–capacitor combination.

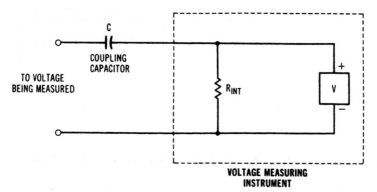

Figure 2.8 A DC-coupled instrument can be AC coupled by adding a capacitor in series with the input.

2.4 RF Probes

Since it is difficult to design an AC voltmeter that has very wide bandwidth, a *radio frequency probe* (RF probe) is often used to enable a DC voltmeter to make high frequency AC measurements. An RF probe usually detects the peak of the AC waveform and converts it into a DC voltage that is then measured by a DC voltmeter. The probe is adjusted so that the DC voltmeter reads RMS volts, even though it is really responding to the peak value. Therefore, the probe is calibrated only for sine waves. Measuring any other waveform will result in errors (similar to the errors resulting from using a peak-reading AC meter). The RF probe will have some finite frequency range over which it will work (100 kHz to 500 MHz, for example). The frequency of the waveform being measured must, of course, fall within this bandwidth. Figure 2.9 shows a typical RF probe circuit.

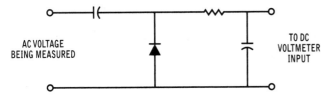

Figure 2.9   A typical RF probe circuit converts a radio frequency AC signal to a DC level readable by a DC voltmeter.

Example 2.2

The RMS value of three waveforms are to be measured: 1) A 60-Hz sine wave, 2) a 100-Hz triangle wave, and 3) a 2-MHz sine wave. The following AC voltmeters are available: A) An average responding meter with 20-Hz to 10-MHz bandwidth and B) a True RMS meter with 5-Hz to 100-kHz bandwidth. Which meters will accurately measure which waveforms?

Meter A can accurately measure only sine waves, since it is an average responding meter (assuming that no correction factors are available). Both sine waves are well within the bandwidth of the meter, so waveforms 1 and 3 can be measured.

Meter B can measure any waveform shape, since it is a True RMS meter, but the 2-MHz sine wave is outside the bandwidth of the meter. So Meter B can measure waveforms 1 and 2.

2.5   Ammeters

The *ideal ammeter* (Figure 2.10) senses the current going through it while maintaining zero volts across its terminals. This implies that the meter must have zero internal resistance and acts like a short circuit when connected to the circuit under test. Recall from Chapter 1 that current measurements are made by breaking the circuit at the point of interest and inserting the instrument such that the current being measured flows through the meter. Since it is connected in series and has zero ohms, it does not affect the circuit under test. An ideal ammeter cannot be achieved in practice, but meters that have very low internal resistance can come close.

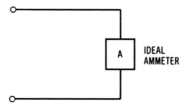

Figure 2.10   The ideal ammeter has zero internal resistance.

A more realistic model of a *real ammeter* (Figure 2.11) includes the internal resistance of the meter. If the internal resistance of the meter is small enough compared with the resistance of the circuit under test, then there will be only a minimal effect on the circuit being measured and the meter will approximate the ideal case.

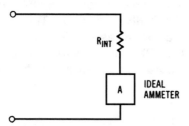

Figure 2.11   A real ammeter has a nonzero (but usually small) internal resistance.

Example 2.3

An ammeter with an internal resistance of 100 Ω is used to measure the current shown in Figure 2.12a. What is the error in the measurement (due to the internal resistance of the ammeter)?

If the ammeter were ideal and had zero internal resistance (Figure 2.12b), the measured current would simply be the value of the voltage source, 10 volts divided by the 2 kΩ resistance.

$$I = 10 \text{ V}/2k = 5 \text{ mA}$$

But with a real ammeter (Figure 2.12b), the internal resistance of the ammeter must be taken into account. Since the meter is in series with the circuit, the resistances add together.

$$I = 10 \text{ V}/(2k + 100) = 4.76 \text{ mA}$$

$$\text{Error} = \text{real} - \text{ideal} = 5 - 4.76 = 0.24 \text{ mA}$$

The error is 4.8% of the ideal value.

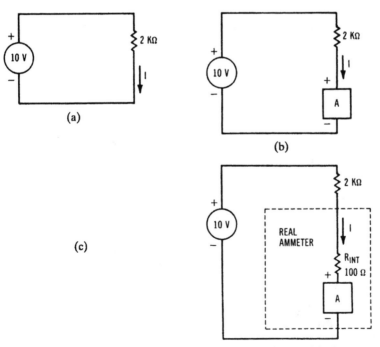

Figure 2.12 An example of current measurement using an ideal ammeter and a real ammeter. a) Circuit to be measured. b) Measurement using an ideal ammeter. c) Measurement using a real ammeter.

2.6   Ammeter Used as a Voltmeter

An ammeter can be configured such that it measures voltage. Many meters, in fact, do use this technique internally to implement the voltmeter function while the actual metering mechanism responds to current. Figure 2.13 shows an ideal ammeter with a series resistance connected. (This series resistor should not be confused with the internal resistance of a real ammeter. If a real ammeter were being considered here, its internal resistance would add in series with R_S.) The current through the meter, I, is given by Ohm's Law:

$$I = V_M/R_S$$

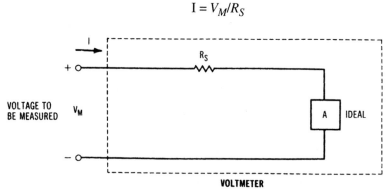

Figure 2.13   An ammeter can be configured to measure voltage by adding a resistor in series.

The current that the meter will read is proportional to the voltage being measured. For simplicity, consider the case where $R_S = 1$ kΩ.

$$I = V_M/1 \text{ k}\Omega$$

If the ammeter was originally calibrated to read mA, then in this configuration it will display V_M in volts. The values in this example were chosen to illustrate the concept easily; in practice, other values for R_S may be used as long as the meter scale is chosen to display the voltage properly. Note that R_S is now the internal resistance of the simulated voltmeter. Since a large (ideally infinite) internal resistance is desired for a voltmeter, R_S is usually chosen to be fairly large and is limited by the sensitivity of the ammeter being used. (If R_S is chosen too large, then very little current will flow through the ammeter, requiring an ammeter capable of measuring very small currents.)

2.7   Voltmeter Used as an Ammeter

A voltmeter can be configured such that it measures current. This technique can be used to make a current measurement even though only a voltmeter is available (or is more convenient). Figure 2.14 shows an ideal voltmeter connected in parallel with R_P. (This parallel resistor should not be confused with the internal resistance of a real voltmeter. If a real voltmeter was shown instead of an ideal voltmeter, its internal resistance would add in parallel with R_P.) Since the ideal voltmeter has no current flowing through it, all of the current, I_M, must be flowing through R_P. Therefore, the voltage across the ideal voltmeter is:

$$V = I_M R_P$$

The voltage displayed by the voltmeter is proportional to the current being measured. Suppose that R_P is equal to 1Ω, then one amp of current through the new ammeter will correspond to a 1-volt reading on the voltmeter. Other values can be used for R_P as long as the scale of the voltmeter is modified to provide the correct current reading. R_P is the internal resistance of the simulated ammeter, so it is desirable to make R_P as small as possible. The value of R_P will be limited by the sensitivity of the voltmeter and the amount of current being measured. For very small values of R_P, very little voltage will appear across it, requiring a very sensitive voltmeter.

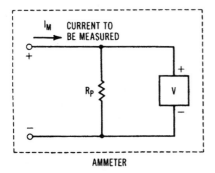

Figure 2.14 A voltmeter can be used to measure current by adding a resistor in parallel.

2.8 Current-Sense Resistor

Another interpretation of Figure 2.14 is that R_P is being used as a *current-sense resistor*. Again, R_P is chosen to be very small so that the circuit being measured will not be disturbed. For example, suppose the current, I_M, in Figure 2.15a is to be measured. Figure 2.15b shows a small 1 Ω current-sense resistor placed in series where the current is to be measured. Since the 1 Ω resistor is very small compared to the 20-kΩ resistor, it will not affect the value of I_M significantly. It does, however, provide a handy place to connect a voltmeter or other voltage measuring instrument so that the current, I_M, can be determined. The current is calculated using the measured voltage (across the current-sense resistor) divided by the value of the current-sense resistor. In this particular example, the voltage across the 20 kΩ resistor could have just as easily been measured and used to determine the current. In many circuits, a resistor is either not available or not conveniently located for current measurement. In those cases, a small current-sense resistor can be added without affecting the operation of the circuit.

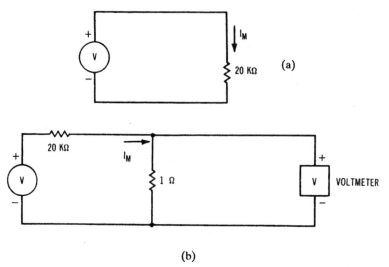

(b)

Figure 2.15   The use of a current-sense resistor is shown. a) The current to be measured flows through the 20 kW resistor. b) A small current-sense resistor is inserted in series and the voltage across it is measured.

Example 2.4

A current-sense resistor is to be used to measure a current that varies between 4 and 6 mA. If a voltmeter with full scale equal to 1 volt is used, what value of resistor will provide maximum sensitivity?

For maximum sensitivity, the voltmeter should read full scale when the current is 6 mA.

$$R = V/I = 1 /0.006 = 166.7 \ \Omega$$

The resulting voltmeter reading in volts must be multiplied by 1/ 166.7 = .006 to determine the current in amps. (Multiply the voltage by 6 to get the current in milliamps.) The circuit under test should be evaluated to see if the current sense resistor will disturb it significantly.

2.8.1   Current Shunt

A *current shunt* is a current-sense resistor designed to measure high values of current. The current shunt attaches to the input of the meter, which is set to measure voltage. The measured current passes through the shunt, producing a voltage for the meter to measure. A typical current shunt can measure up to 30 amperes, with a conversion coefficient of 1 mV/A. (This implies that the shunt's resistance is 0.001 Ω.)

The maximum current rating on the shunt must not be exceeded. With such high current levels, excessive power dissipated in the shunt can destroy it.

2.9 AC Ammeter

The AC ammeter is very similar to the DC ammeter except that it measures the current of AC (not DC) waveforms. The same ideal ammeter and real ammeter circuit models shown in Figure 2.10 and Figure 2.11 apply to the AC ammeter. Again, the ideal ammeter does not affect the circuit under test, due to its zero internal resistance, but a real ammeter will have some small resistance.

All of the considerations discussed under AC voltmeters (average vs. true RMS responding meters, bandwidth, AC vs. DC coupling) apply to AC ammeters as well. The reader simply needs to mentally substitute "current" for every occurrence of "voltage."

2.10 Ohmmeters

An ohmmeter measures the resistance of the device or circuit connected to its input. There are many different meter configurations that can be used to implement an ohmmeter. Several of the more common configurations are described later. It is usually not necessary to know the internal operation of the ohmmeter as long as the operating manual is followed correctly. What is important is that certain general principles be understood when making resistance measurements.

Since resistance measurements are derived from voltage and current measurements, ohmmeters supply their own stimulus (voltage or current) to the device being measured. This allows individual resistors to be measured without being part of a functioning circuit. This also means that if the resistance is part of a circuit, that circuit must have other sources of voltage (or current) removed from it. Power supplies and batteries must be turned off or disconnected from the circuit. If not, currents or voltages induced will cause the ohmmeter reading to be incorrect and, if large enough, may damage the meter. (Compare this to voltage and current measurements where the circuit must be powered up to get any meaningful readings.)

Ohmmeters operate using DC voltages and currents, therefore, the DC resistance of the device is measured (not AC impedance). Attempts to measure AC impedance with an ohmmeter result in inaccurate and frustrating readings. The classic example is the person who tries to verify the impedance of a loudspeaker, specified as 8 ohms, using an ohmmeter. The ohmmeter will actually measure the DC resistance of the speaker, which is generally just the resistance present in the voice coil winding. This reading could be most any value and does not indicate the AC impedance of the speaker. Also, measurements on components other than resistors, such as diodes, transistors, and capacitors may be misleading (although diode measurements are discussed later).

A zero resistance adjustment is often provided in an ohmmeter. This adjustment compensates for some of the measurement drift internal to the meter as well as things external to the meter such as test lead resistance. The usual procedure is to short the test leads together and adjust the meter until it reads zero. This procedure compensates for the resistance of the test leads, removing their effect from the measurement. If the test leads are removed or changed for a particular measurement, the meter should be zeroed again. In some meters it is also necessary to zero the meter on every different range setting. Instead of a zero adjustment, some meters have an infinity adjustment. In that case, the test leads are held apart (open circuited) and the meter is adjusted to read infinity. Other meters have both zero and infinity adjustments, which must be adjusted separately.

Resistance exists between two points, very similar to voltage. When an ohmmeter is connected to two points (for example, two ends of a resistor being measured), all of the circuit paths from one point to the other will be measured. If there are two resistors in parallel between the two points, the parallel combination will be measured. If there are two resistors in series between the two measurement points, then the series combination will be measured. This must be carefully considered when measuring resistors that are part of a circuit. Simply connecting an ohmmeter across a resistor in a circuit does not necessarily measure only that resistor, since there may be other components connected in parallel. Either these components must be accounted for in the resulting measurement or one end of the resistor being measured must be disconnected to guarantee that no other device is connected to it.

The human body has a resistance that varies considerably and affects high resistance measurements. The conducting part of the test leads should not be touched by the user when making resistance measurements above about 10 kΩ, otherwise body resistance may affect the measurement by appearing in parallel with the resistor being measured.

2.11   Voltmeter–Ammeter Method

One obvious way to measure resistance is to connect a voltmeter and ammeter, as shown in Figure 2.16. A voltage source and resistor, V_S and R_S, cause a current to flow through the resistance being measured (R_X). The current through and the voltage across R_X are measured by the voltmeter and ammeter, and the value of R_X can be computed using Ohm's Law, $R_X = V_M/I_M$.

This method is useful for making in-circuit tests when a voltmeter and ammeter are available. In such a case, V_S and R_S are not used, since the circuit operation presumably causes some current to flow through R_X, allowing the measurement to be performed. This method is not normally used in ohmmeters, since it requires both an ammeter and a voltmeter to be implemented inside the ohmmeter.

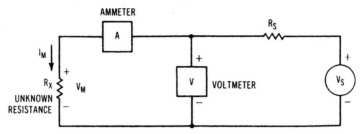

Figure 2.16   The voltmeter–ammeter method measures the voltage and current across the unknown resistance. The value of the resistance is then computed using Ohm's Law.

2.12   Series Ohmmeter

An ohmmeter implementation requiring only 1 meter is shown in Figure 2.17. R_X is the resistance being measured, while R_1 and V_S are known values internal to the ohmmeter. The resulting current through R_X can be computed as follows :

$$I_X = V_S / (R_1 + R_X)$$

Note that I_X is *not* directly proportional to the value of R_X. Therefore, since the ammeter reading is used to indicate resistance, the ammeter does not have a simple linear scale but looks something like Figure 2.18 (assuming an analog meter is used). A full-scale current reading indicates zero resistance while a zero current reading indicates infinite resistance. (The previous statement can be confirmed with a quick review of the circuit in Figure 2.17.)

R_1 is typically made variable to allow for precise calibration at the $R_X = 0$ point. Normally, the test leads of the ohmmeter are shorted together and R_1 is adjusted until the meter reads zero ohms. This adjustment compensates for changes in V_S (which may be a battery) and for test lead resistance.

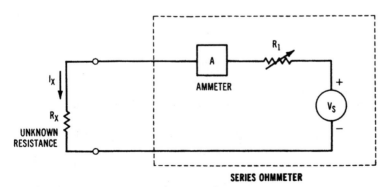

SERIES OHMMETER

Figure 2.17   The series ohmmeter measures the unknown resistance using only 1 ammeter.

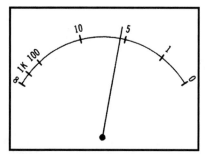

Figure 2.18 The meter scale for the series ohmmeter. Notice that it is not a simple linear scale.

2.13 Current–Source Method

Another method used to implement an ohmmeter is to pass a known current through the unknown resistance and measure the resulting voltage (Figure 2.19). The current source, I_S, causes a fixed current to flow, all of which passes through R_X (since an ideal voltmeter has infinite resistance). Since $V_M = I_S R_X$, the measured voltage is directly proportional to the unknown resistance, so this method results in a linear relationship between the meter reading and the unknown resistance.

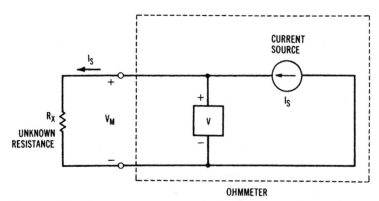

Figure 2.19 The current source method for implementing an ohmmeter passes a known current through the resistor being measured. The voltmeter is then used to measure the voltage across the resistor, which is proportional to the resistance.

2.14 4-Wire Ohms Measurements

When measuring very small resistances, the resistance of the meter test leads may introduce a significant error. Figure 2.20a shows the circuit model of a current–source method ohmmeter with the resistance of the test leads included. The current I_S flows through these lead resistances and produces a voltage across each of them. Since the voltmeter will include these voltages in its measurement, the lead resistances add to the unknown resistance, R_X, producing an error. Since high-quality test leads have resistances of less than an ohm, this is only a problem when measuring small resistances.

The *4-wire ohms* technique removes this error by using two additional test leads. Figure 2.20b shows the test configuration for the 4-wire ohms measurement. The current I_S flows through a separate set of leads to the unknown resistance. These leads also have some finite resistance in them but the current source will maintain a constant current to the unknown resistance. The other two test leads are used by the voltmeter to monitor the voltage across the unknown resistance. Since there is no current flowing through these two test leads, there is no voltage drop across them. Therefore, the voltmeter can accurately sense the voltage across R_X and since the current through the resistor is known, the resistance can be determined. The key to the 4-wire ohms technique is keeping the current out of the voltmeter leads. (The internal resistance of the voltmeter must be very high.)

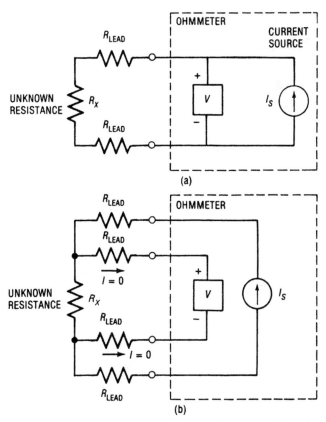

Figure 2.20 a) Equivalent circuit showing error effect due to test-lead resistance. b) Equivalent circuit for the 4-wire ohms measurement with test-lead error eliminated.

2.15 Multimeters

The most popular type of meter, available in both analog and digital form, is the *multimeter*. A multimeter is simply a voltmeter, ammeter, and ohmmeter combined into one instrument. Internally, either an ammeter or voltmeter is implemented, which is then used to perform the other functions. Other names given to the multimeter are *volt-ohm-milliammeter* or *VOM* (usually an analog meter) and *digital multimeter* or *DMM* (always a digital meter). An example of a handheld digital multimeter is shown in Figure 2.21 and a summary of its specifications are listed in Table 2.1.

Figure 2.21   A handheld digital multimeter from Fluke Corporation. (Reprinted with permission.)

Table 2.1   Abbreviated Specifications of a Handheld Digital Multimeter

Specification	Value
Basic DC accuracy	0.025%
Basic AC accuracy	0.4%
Resolution	50,000 counts ($4\frac{1}{2}$ digits)
AC bandwidth	100 kHz
DC voltage ranges	50 mV to 1,000 V
DC current ranges	500 μA to 10 A
AC voltage ranges	50 mV to 1,000 V
AC current ranges	500 μA to 10 A
Resistance ranges	500 Ω to 500 MΩ

2.16   Meter Range

The techniques discussed for implementing voltmeters, ammeters, and ohmmeters are useful over a limited measurement range. Circuit techniques are used to expand or contract this measurement range to supply the user with different measurement settings. Range selection may be implemented by switching different resistor values into the metering circuit or by switching in an amplifier at the input of the meter. A multimeter functioning as a voltmeter, for example, might have 100-mV, 1-volt, 10-volt, and 100-volt ranges. For maximum accuracy, the lowest range larger than the voltage to be measured is used. A larger range will work, but with reduced accuracy. A lower range will cause the meter to be overloaded, possibly damaging it. If the approximate value of the voltage being measured is unknown, then starting with the maximum range is recommended.

2.16.1   Autorange

Meters with an autorange function automatically choose the best range for a given measurement. Autoranging meters essentially duplicate the actions of an experienced instrument user. If the meter is currently being overloaded, a higher range is selected. If the measured value is small enough that a lower range could be used (with greater accuracy), then a lower range is selected. Autoranging meters usually let the user disable the autorange feature and select the range directly. This is convenient when a particular range is required or the correct range is already known. Since it takes time for the meter to autorange, repetitive measurements near the same value may proceed faster if the range is selected directly.

2.17 Other Multimeter Functions

Other features are often included in the modern digital multimeter along with the basic voltage, current, and resistance measurements. These features build on the versatility of the DMM, providing additional convenience and utility.

2.17.1 Continuity Indicator

When tracing out wires in a circuit, an ohmmeter can be used to determine whether or not a connection exists between two points. If the resistance value is very low (typically less than a few ohms), there is circuit continuity. If the resistance value is high, then no connection exists. Since the actual resistance measured is not all that critical for many continuity checks, some meters have an audible indicator that can be set to beep when the resistance is less than some value (say, 10 ohms). Thus, the user does not have to turn and look at the meter, but instead can focus on the circuit being tested.

2.17.2 Diode Test

Another common feature is a special diode test mode. A known current is forced through the diode and the forward voltage drop across the diode is displayed. The current value is typically between 0.5 and 1 mA, large enough to switch most diodes on, but small enough so that sensitive diodes will not be damaged. Since a diode conducts in only one direction, proper polarity must be maintained when making the measurement. In some meters, the diode test mode is just a particular range of the ohmmeter function set up so the display reads out the voltage drop. Diode leakage (the flow of current in the reverse direction) may be tested by reversing the polarity of the leads.

2.17.3 Frequency Counter

Many DMMs include a frequency counter feature that measures the frequency of an AC signal. Like frequency counters discussed in Chapter 6, reciprocal counting techniques are used to determine the frequency. Frequency measurements in DMMs are generally limited to the AC bandwidth of the meter, usually a few hundred kilohertz or less. This implies that the frequency measurement function is useful mostly for checking power line, audio, and other low-frequency signals.

2.17.4 Minimum, Maximum, Average Readout

For situations when the meter reading varies over time, a meter may include *minimum,* *maximum,* and *average* functions. When one of these functions are invoked, the meter monitors the measured reading and keeps track of the minimum, maximum, or average of the reading.

The average function can be used to remove minor variations in the reading (perhaps due to electrical noise) to produce a better estimate of the true voltage reading. The averaging function may also have the benefit of increasing the resolution of the measurement.

The minimum or maximum function can be used to catch infrequent surges or dropouts in voltage. For example, a meter could be connected to the power line to capture the maximum or minimum line voltage over a period of several hours.

2.17.5 Capacitance Measurement

Basic capacitance measurement is sometimes included in DMMs. Such a capacitance measurement is performed at low frequencies with a typical accuracy of 1% to 2%. This function is useful for checking the value of a component.

2.17.6 Temperature Measurement

Some DMMs are set up to perform temperature measurements, requiring external temperature sensor probes. There are two basic types of temperature probes commonly used. *Thermistor probes* cover a temperature range of approximately –80°C to 150°C (–112°F to 302°F). Measurement accuracy is usually better than 1°C. *Thermocouple probes* use Type K thermocouples that operate over a wider temperature range, –260°C to 260°C (–436°F to 500°F), with a measurement accuracy of several degrees C.

2.18 Specifications

The exact method of specifying meter performance varies with both instrument model and manufacturer, but generalities do exist. Voltmeter, ammeter, and ohmmeter accuracy are usually specified as percent of full scale (i.e., percent of the range), as percent of the reading, or both. Digital meters usually add to this an uncertainty of several digital counts.

Each different function of a meter (AC voltmeter, DC voltmeter, ohmmeter, etc.) as well as each range of each function may have separate specifications. It is important to examine the manufacturer's specifications carefully to understand the expected measurement accuracy.

Example 2.5

A $3\frac{1}{2}$-digit digital multimeter used on the 20-volt range (maximum reading is 19.99 volts) has an accuracy specification of ± (0.75% of reading + 2 counts). What is the maximum error if the input voltage is 12 volts?

One count is the smallest change that the meter can display (resolution), in this case, 0.01 volts. (With $3\frac{1}{2}$ digits the meter can read 12.00, 12.01, 12.02, etc.).

Error = ± (0.75% × 12 volts) + (2 × 0.01)

Error = ± (0.09 + 0.02)

Error = ± 0.11 volt

So the meter reading could be as low as 11.89 or as high as 12.11 volts. (This analysis does not include other sources of error such as the loading effect described in Chapter 1.)

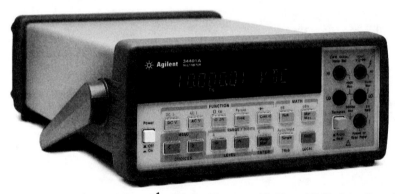

Figure 2.22   A bench top $6\frac{1}{2}$-digit multimeter with True RMS detection. (Photo: Agilent Technologies. Reprinted with permission.)

There are a variety of multimeters available in the marketplace, with a corresponding wide range of measurement capability as well as price. Generally, increased accuracy, resolution, and features causes a corresponding increase in cost. On the low end are small, portable analog VOMs. Also small and portable, but with improved accuracy and a digital display, is a compact digital multimeter such as the model shown in Figure 2.21. Figure 2.22 shows a $6\frac{1}{2}$ digit digital multimeter intended for bench use. The abbreviated list of specifications for this meter are listed in Table 2.2.

Table 2.2   Abbreviated Specifications of a $6\frac{1}{2}$-Digit Multimeter

Specification	Value
DC volts	Ranges: 0.1, 1, 10, 100, 1000 V
	Accuracy: ± (0.005% of reading + 0.0035% of range)
AC volts	Ranges: 0.1, 1, 10, 750 V
	Accuracy: ± (0.06% of reading + 0.03% of range), 10 Hz to 20 kHz
DC current	Ranges: 0.01, 0.1, 1, 3 A
	Accuracy: ± (0.1% of reading + 0.01% of range)
Resistance	Ranges: 100 Ω, 1 kΩ, 10 kΩ, 100 kΩ, 1 MΩ, 10 MΩ
	Accuracy: ± (0.01% of reading + 0.001% of range)

2.19   References

"ABCs of DMMs," Fluke Corporation, Application Note, March 1999.

"Agilent 34401A Multimeter User's Guide," Agilent Technologies, Publication No. 34401-90004, February 1996.

Coombs, Clyde F., Jr. *Electronic Instrument Handbook*, 3rd ed., New York: McGraw-Hill, 1999.

"Fluke 187/189 True-RMS Digital Multimeter—Extended Specifications," Fluke Corporation, Document #2153, August 2000.

Gottlieb, Irving M. *Basic Electronic Test Procedures*, 2nd ed., Blue Ridge Summit, PA: Tab Books, 1985.

Gottlieb, Irving M. *Basic Electronic Test Procedures*, 2nd ed., Blue Ridge Summit, PA: Tab Books, 1985.

Oliver, Bernard M., and John M. Cage. *Electronic Measurements and Instrumentation*. New York: McGraw-Hill, 1971.

Wedlock, Bruce D., and James K. Roberge. *Electronic Components and Measurements*. Englewood Cliffs, NJ: Prentice Hall,1969.

Signal Sources

Most of the instruments discussed in this book measure some sort of electrical parameter, usually voltage or current. In this chapter, we will discuss a class of instruments that do not measure any-thing—at least not by themselves. Instead they provide the signals that stimulate the circuit or device being tested, so that other instruments may measure the electrical parameters of interest.

We will use the term *signal source* to mean an instrument that supplies a known signal to a circuit being tested. The characteristics of that signal will vary depending on the partic-ular type of signal source. Some sources produce only a single frequency sine wave. Other, more advanced instruments, provide signals such as triangle waves, square waves, and pulses and can sweep the frequency of the waveform as well as modulate the signal. Although various distinct categories of signal sources will be discussed, many instruments have characteristics that transcend several categories.

3.1 Circuit Model

Although the actual circuitry of a signal source may be very complex, its electrical characteristics can be modeled with a Thevenin circuit (Figure 3.1). V_S is the open circuit voltage of the source and R_S is the output resistance. V_S is not limited to being a sine wave, but can be any waveform that a signal source is capable of producing. This includes square and triangle waves, pulse trains, and modulated signals. The output resistance is typically one of several common values: 50, 75, and 600 ohms.

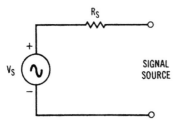

Figure 3.1   The circuit model for a signal source is a voltage source with a series resistance.

Example 3.1

A sine wave source with a 50 Ω output resistance is supplying 3 volts RMS across a 50 Ω load resistor. How much power will be supplied to the load if the 50 Ω resistor is replaced by a 75 Ω resistor?

Figure 3.2a shows the circuit model for the source connected to a 50 Ω load. Since $V_L = V_S \, 50/(50 + 50)$ by the voltage divider equation,

$$V_S = 2 \, V_L = 6 \text{ volts RMS}$$

Figure 3.2b shows the situation with the 75-Ω load resistor connected. In this case, $V_L = (6)75/(75 + 50) = 3.6$ volts RMS

The power delivered to the load is

$$P = V^2_L/R = 3.6^2/75 = 0.173 \text{ watts}$$

The power can be expressed in dBm, $P = 10 \log (0.173/0.001) = 22.4$ dBm

---

3.2   Floating and Grounded Outputs

The output of a signal source may be floating (Figure 3.3a) or grounded (Figure 3.3b). If the output is floating, then neither of the two output terminals is connected to the instrument chassis. If the output is grounded, then one of the output terminals is connected to the instrument chassis, which in turn is connected to the power line safety ground (via the ground connection on the three-terminal AC power plug). Some instruments offer a floating output with a switch or jumper built in to allow convenient grounding of the output if desired.

Floating outputs are more versatile, but are more difficult to design and manufacture, especially for high frequencies. Many circuits have one side of their input connected to ground anyway, so a grounded output source can often be used with no problem. In some cases, however, it is desirable to have neither side of the signal source connected to ground. For example, the source may be driving a circuit at a point, which has both input connec-

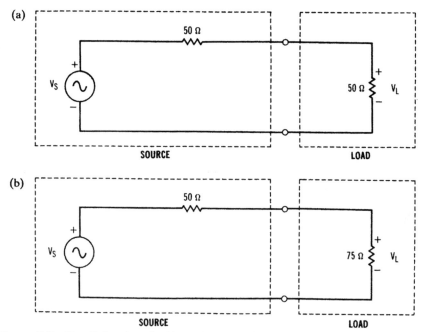

Figure 3.2 Circuit drawings for Example 3.1. a) A signal source with a 50-ohm load resistor. b) The same signal source with a 75-ohm load resistor.

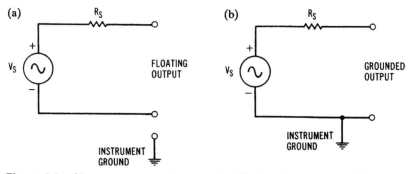

Figure 3.3 Signal source circuit model with (a) a floating output and (b) a grounded output.

tions at several volts above ground. If a grounded source were connected, one of these input points would then be grounded and the operation of the circuit may be disrupted.

3.3 Sine Wave Sources

As the name implies, sine wave sources are capable of supplying only sine wave signals. These sources are economical instruments that generate low distortion signals ranging in frequency from a few Hz to about 1 MHz. Internally, the technology used is usually a free-running oscillator operating at the same frequency as the desired output frequency. This technology is inexpensive, but is limited in frequency range. Sine wave oscillators are used mainly for audio frequency work, and may be optimized for producing very low distortion signals.

Figure 3.4 shows a simplified block diagram of a sine wave source. It consists of a free-running oscillator followed by an amplifier to boost the signal level and an attenuator to vary the output level. The free-running oscillator frequency is usually controlled by a variable capacitor and/or resistor. Figure 3.5 is an example of a sine wave source, this one having an auxiliary square wave function.

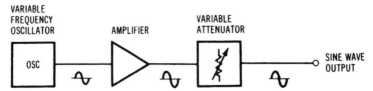

Figure 3.4   Simplified block diagram of a sine wave source shows a free-running oscillator followed by an amplifier and attenuator.

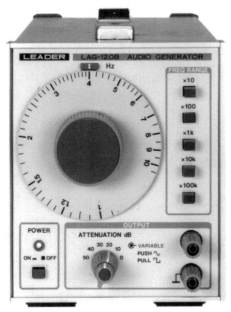

Figure 3.5   A sine wave oscillator that can also output a square wave. (Photo courtesy of Leader Instruments Corporation.)

3.4 Imperfections in Sine Wave Sources

Signal sources do not produce absolutely perfect, undistorted signals, but instead will have a variety of imperfections in the signal. Different applications require different levels of performance in the signal source, so the user must understand these imperfections and how they relate to the measurement.

A pure, undistorted sine wave would appear in the frequency domain as shown in Figure 3.6a. Recall from Chapter 1 that a sine wave has energy at only one precise frequency with no harmonics or other frequency components. This perfect sine wave does not vary in amplitude or in frequency.

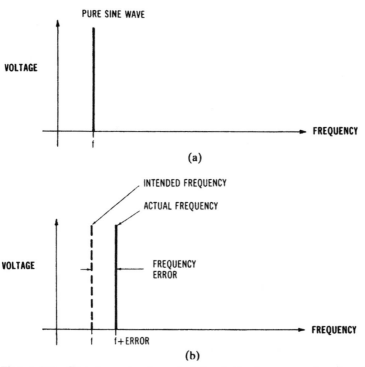

Figure 3.6 Signal source characteristics in the frequency domain. a) A pure sine wave appears as a single frequency component. b) A frequency error shows up as a shift along the horizontal axis. c) An amplitude error causes a change in the height of the spectral line. d) Harmonics are additional frequency components at multiples of the original frequency. e) Spurious responses are unwanted frequency components at other non-harmonic frequencies. f) Close-in sidebands can be caused by residual amplitude, frequency, or phase modulation.

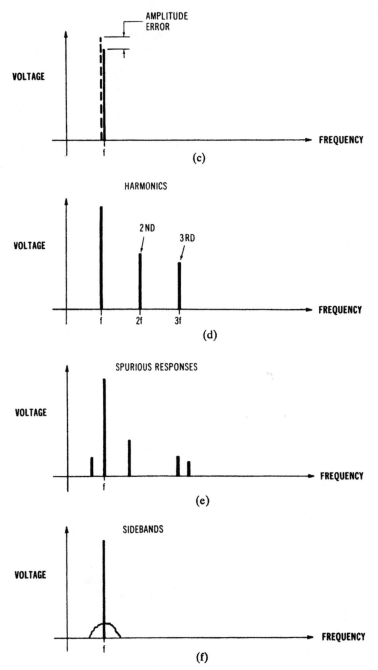

Figure 3.6   (Continued)

3.4.1   Frequency Accuracy

The *frequency accuracy* of a signal source determines how precisely the actual frequency of the waveform matches the frequency setting on the front panel of the instrument. Frequency accuracy is usually specified in percent. A frequency error shows up in the frequency domain as a horizontal shift of the sine wave's vertical line (Figure 3.6b).

3.4.2   Frequency Stability

The sine wave from a signal source may vary somewhat in frequency depending on the source's *frequency stability*. This results in a frequency error that varies with time. In the frequency domain, one can imagine an unstable signal wandering slightly back and forth along the frequency axis. This wandering is typically rather slow, but in extreme cases a source could be so unstable that the resulting waveform has noticeable frequency modulation sidebands. There is a fine and arbitrary distinction between frequency instability and actual frequency modulation. Usually, a slow frequency variation (less than 1 Hz) is classified as frequency instability and a fast variation is classified as residual frequency modulation. Frequency stability is usually specified in percent or parts per million (ppm) and may be valid only after a specified warm-up time.

Frequency stability and frequency accuracy are related but are not the same. If a source had perfect frequency stability but poor frequency accuracy, then the frequency would remain constant, but at the wrong frequency. The frequency inaccuracy of such a source could be compensated for by tuning it to the proper frequency while monitoring it with a more accurate instrument, such as a frequency counter. On the other hand, a source with poor frequency stability cannot achieve good frequency accuracy, at least not for any length of time. Such a source could be set to the correct frequency for an instant but would quickly drift away from it.

3.4.3   Amplitude Accuracy

Amplitude accuracy specifies how precisely the actual amplitude of the waveform matches the source's control setting. An amplitude error in the frequency domain will result in a vertical line at the proper frequency, but with an incorrect height (amplitude), as shown in Figure 3.6c.

Some instruments have an *amplitude flatness* specification that describes how much the output amplitude varies over a specific frequency range, usually the maximum range of the instrument. Once the user sets the source's amplitude (perhaps by monitoring it with another instrument), the amplitude will remain within the limits of the flatness specification even though the frequency is changed.

3.4.4   Distortion

The waveform may have zero frequency error and zero amplitude error, but have considerable *distortion*. Distortion is any imperfection in the shape of the waveform as compared against the ideal signal. In the sine wave case, distortion shows up in the frequency domain as undesirable harmonics, which is referred to as *harmonic distortion*. Harmonic distortion is shown in Figure 3.6d with the original pure sine wave now accompanied by several frequency components at integer multiples of the original frequency. The maximum amplitude of the harmonics may be specified as a percent of the fundamental or in dB relative to the fundamental. All of the harmonics may be lumped into one distortion number called *total harmonic distortion* (THD) and specified in percent.

3.4.5   Spurious Responses

A signal source may produce low-level frequency components that are not harmonically related to the output frequency. These components are called *spurious responses* and are often the result of the particular signal-generation technique used inside the instrument. Alternatively, they may result from the AC power line frequency appearing either directly or as modulation on the carrier. Spurious responses may appear at any frequency and may move when the source's frequency changes. Figure 3.6e shows the effect of spurious responses in the frequency domain. The maximum value of this type of response is usually specified at an absolute level or in dB relative to the desired signal (or both).

3.4.6   Close-in Sidebands

If the sine wave were absolutely pure, in the frequency domain it would have an infinitely thin spectral line, indicating that all of the signal's energy is at exactly one frequency. In reality, the sine wave's frequency response may spread out slightly, often causing a pedestal effect, as shown in Figure 3.6f. These imperfections usually appear as noise sidebands very close to fundamental and are thought of and specified in a variety of ways. Some instruments specify these sidebands as *residual modulation* (AM or FM), while others refer to the phenomenon as *phase noise*. These different mechanisms are not exactly equivalent, but they all produce a noise-like response close to the output frequency.

3.5   Function Generators

Function generators are the most widely used general-purpose signal source. They are capable of supplying a variety of waveforms including sine waves, square waves, and triangle waves (Figure 3.7). Depending on the particular instrument, a function generator may also provide additional waveforms, modulation, and swept-frequency capability.

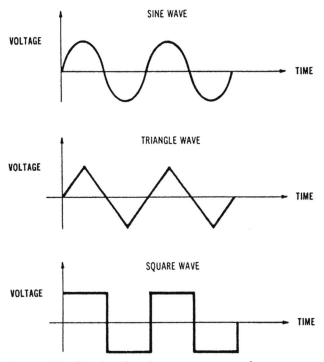

Figure 3.7 Standard function generator waveforms.

The function generator may use technology similar to the sine wave oscillator. A simplified block diagram is shown in Figure 3.8. One of the required waveforms is produced by a free-running oscillator and then conversion circuits derive the other waveforms from the original. In Figure 3.8, a triangle wave is generated by the oscillator. A square wave is

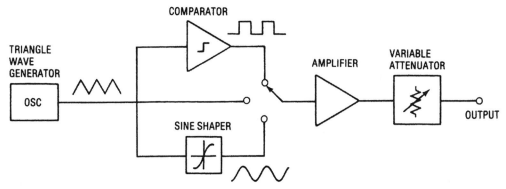

Figure 3.8 Simplified block diagram of a function generator using a free-running oscillator, which generates a triangle wave. The triangle wave is converted to a square wave or sine wave to obtain the standard function generator waveforms.

derived from the triangle wave by running the triangular wave through a comparator. (The comparator may actually be part of the triangle wave oscillator.) A sine wave is derived from the triangle wave by passing it through a wave-shaping circuit (called a *sine shaper*). The desired waveform is selected, amplified, and output through a variable attenuator.

Function generators typically provide a DC offset adjustment that allows the user to add a positive or negative DC level to the generator output. Figure 3.9 shows how adding varying amounts of DC to a square wave can produce different waveforms. The DC offset can also be used to control the DC bias level into the circuit under test.

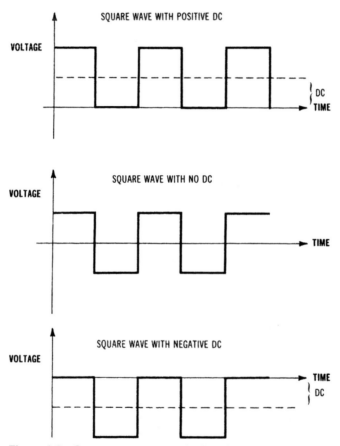

Figure 3.9   Square wave output of a function generator with varying amounts of DC offset added.

As the technology has improved, the function generator has largely replaced the sine wave-only type source, since it offers sine wave capability along with the other waveforms to provide much greater flexibility. Function generators can be used for audio sine wave testing just as the sine wave-only generator can. In addition, the square wave output can be used as a clock for digital circuits. Some function generators include a fixed TTL-compatible output for such an application. On others, square wave output and DC offset adjustment can be used to generate valid logic levels. The triangle wave can be used in situations where a ramp-like voltage is needed. Figure 3.10 shows an example of a modern function generator.

Figure 3.10 A 15-MHz function generator with arbitrary waveform capability. (Photo: Agilent Technologies. Reprinted with permission)

3.5.1   Arbitrary Waveform Generators

The basic function generator waveforms are useful for many applications and have been designed into common test procedures. However, it is useful to be able to create more complex, even arbitrary waveforms. A special class of source has emerged, the *arbitrary waveform generator* (AWG), which has the ability to generate arbitrary waveforms.

AWGs use a technique called *direct digital synthesis* (DDS) to generate waveforms. The desired waveform is stored as digital numbers that drive a *digital-to-analog converter* (DAC) that converts the digital representation into an analog waveform (Figure 3.11). The DAC is followed by a low-pass filter that removes unwanted high-frequency content ("aliases") from the synthesized waveform and an amplifier drives the AWG output. Figure 3.12 shows how a sine wave is represented as digital codes, which are used to drive the DAC. In the figure, the range of DAC codes is from 0 to 4,096, consistent with the use of a 12-bit DAC.

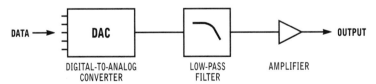

Figure 3.11 Direct digital synthesis (DDS) is accomplished by using a digital-to-analog converter (DAC), which converts a stored digital waveform into its analog form.

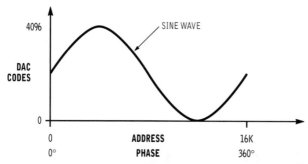

Figure 3.12   The input to the DAC is a series of digital values or codes that represent the desired waveform.

A more complete AWG block diagram is shown in Figure 3.13a. Digital data representing the desired waveform is stored in digital memory (waveform memory). An address counter sequences through the addresses of the memory, causing the data in the memory to sequentially appear at the digital port of a DAC. The DAC produces a voltage proportional to the digital data supplied to it. Thus, as the address cycles through memory, the waveform data is transformed into a voltage waveform at the output of the DAC. An oscillator is used to clock the address counter at a variable rate, to allow the waveform period and frequency to be varied. The faster the address counter is clocked, the faster the memory is cycled through and the higher the waveform frequency. The output of the DAC goes through a low-pass filter, followed by an amplifier and an attenuator.

This particular type of AWG block diagram is sometimes called a "point-per-clock" system, since data to the DAC changes on each clock cycle and data is output sequentially from the waveform memory. That is, the address counter increments through each memory location using all data points in the waveform memory in a sequential manner.

A more complex block diagram is shown in Figure 3.13b. In this system, the address counter is replaced by a *phase register, adder,* and *phase increment register* (PIR). The phase register supplies the address to the waveform memory, just like the address counter in the point-per-clock system. However, the value loaded into the PIR controls the change of the phase register that occurs on each clock cycle. If the PIR is set to 1, then the phase register increments through all addresses and operation is identical to the address counter in the point-per-clock system.

The phase register contains more bits than required by the waveform memory address. For example, the waveform memory might be 16 k samples, requiring an address that is 14 bits in length. The phase register might be 48 bits wide, with only the upper 14 bits actually used for the memory address. The benefit of this approach is that the phase register can be incremented in much finer resolution, with steps that are smaller than one address change on the waveform memory. This provides much finer control over the frequency of the waveform.

The phase register gets its name from the fact that it directly corresponds to the phase of the waveform (refer back to Figure 3.12). When the phase register is zero, the phase of the waveform is zero. When the phase register is one-fourth of the total memory size, the waveform phase is 90 degrees. When the phase register is one-half of the total memory size, the waveform phase is 180 degrees. The phase increment register controls the rate of change in the phase register, which is proportional to the frequency of the signal. If the value of the phase increment register is small, the phase register changes slowly on each clock cycle and the corresponding output frequency is low. If the phase increment register is programmed with a large value, the phase register changes rapidly with a higher waveform frequency as the result.

Since most modern function generators operate under microprocessor control, dynamic control of the phase register and phase increment register can support a number of key features. For example, programming of the phase register can be used to implement phase modulation of the waveform. Altering the value of the PIR can result in frequency shift keying or produce a linear frequency sweep.

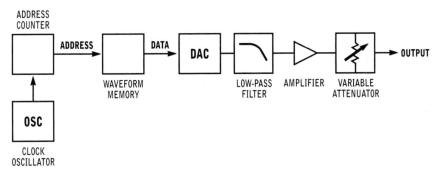

Figure 3.13a   The point-per-clock arbitrary waveform generator block diagram.

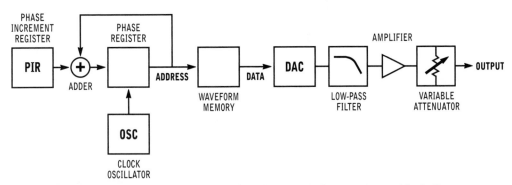

Figure 3.13b   The phase-increment-register arbitrary waveform generator block diagram.

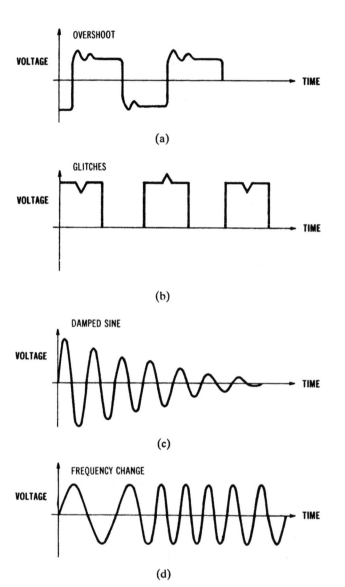

Figure 3.14 Examples of waveforms produced by an arbitrary waveform generator. a) Square wave with overshoot. b) Digital signals with glitches. c) Damped (decaying) sine wave. d) Sine wave with abrupt frequency change.

3.5.2 Arbitrary Waveforms

The application of arbitrary waveform generators is limited only by the number of different signals present in the world. Any waveform that can be stored in digital form and loaded into the waveform memory can be generated. Of course, the arbitrary generator is still limited by the same bandwidth constraints that limit other instruments. Also, the DAC performance, the clock rate, and the size of the waveform memory will limit the frequency range of the generator. The actual implementation of the arbitrary waveform generator varies from compact, stand-alone instruments to generators that are dependent on an external computer to load the waveform data. Many generators supply a "waveform calculator," which allows the user to define the arbitrary waveform easily without manually entering each waveform point.

An arbitrary waveform generator can be used in a variety of applications where previously it was difficult or impossible to generate appropriate waveforms. Of course, it can also be used to generate any of the waveforms previously discussed (square, triangle, pulse), but the real utility of the generator is in simulating more complex signals. Figure 3.14 shows some examples of waveforms that can be produced using an arbitrary waveform generator. Imperfections in signals (such as overshoot in a square wave and glitches in a digital signal) can be simulated in a controlled manner to determine how sensitive a circuit is to such problems. Also, transient signals such as a damped sine wave can be produced. Another advantage of the arbitrary waveform generator is the ability to quickly and precisely change frequency (frequency hop).

As digital technology and DAC technology have improved, the DDS architecture has become much more cost effective than traditional analog approaches. Most modern function generators use DDS and include arbitrary waveform capability as a standard feature, delivering more overall capability to the user as a direct result of the DDS architecture.

3.5.3 Modulation (AM, FM)

Many modern function generators include amplitude modulation (AM) and frequency modulation (FM) as standard features. These features require a *modulating signal* to be applied to a *carrier signal*, as described in Chapter 1. Usually the carrier signal is produced by the main waveform generation circuits of the generator. The modulating signal may be provided by an internal modulation source or another signal source may be required to supply the signal externally.

3.5.4 Burst

Burst mode or *burst modulation* is a powerful feature that allows the generator to output a specific number of cycles of a waveform. For example, a function generator can be programmed to output 3 cycles of a sine wave, as shown in Figure 3.15a. The burst can be triggered by the press of a front panel key ("single trigger") or by an external signal applied to the *external trigger* input. The function generator can be programmed to automatically repeat this 3-cycle burst at a specific rate, called the *burst rate*, resulting in the waveform shown in Figure 3.15b. Burst mode usually applies to all of the waveforms available in the function generator, including arbitrary waveforms. Burst can be combined with different waveform shapes and the external trigger, resulting in powerful waveform generation.

a)

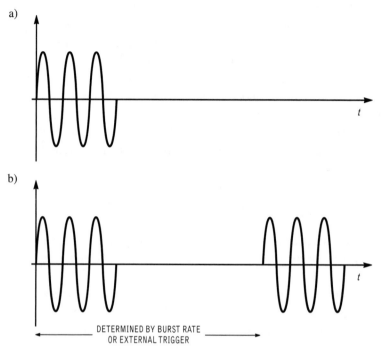

b)

Figure 3.15   a) A single burst of 3 cycles of a sinewave. b) A repeating burst of 3 cycles of a sinewave.

3.5.5   Frequency Shift Keying

Frequency shift keying (FSK) occurs when a function generator is set to instantaneously shift between two frequencies specified by the user (Figure 3.16). The generator may be set to shift (or "key") automatically at some specified rate or an external input may be used to control the shifting.

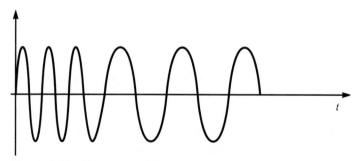

Figure 3.16   Frequency shift keying (FSK) causes the generator to instantly shift between two frequencies.

3.5.6   Frequency Sweep

Many function generators have the ability to sweep their output frequency (see Figure 3.17). This swept frequency capability is useful for testing circuits over a wide frequency range in a short amount of time. A specialized source, called a *sweep generator*, is specifically designed for frequency sweep applications.

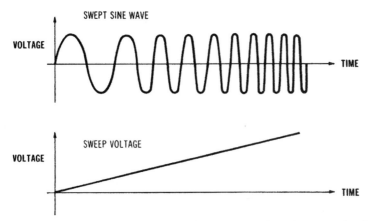

Figure 3.17   The output of a sweeping generator is a sine wave that increases in frequency. The sweep voltage is proportional to the output frequency.

Sweep generators generally sweep in frequency in a linear manner but may also offer a logarithmic frequency sweep. A sweep voltage output that is proportional to the frequency during a sweep is usually provided, which can be used to drive other instruments. In particular, this output can be used with an oscilloscope configured for a frequency response test (see Chapter 5). (The sweep voltage is also called the X-drive output, since it is used to drive the X-axis of the oscilloscope display.) Figure 3.17 shows a sine wave being swept in frequency along with its corresponding sweep voltage.

Figure 3.18 shows a conceptual block diagram of a sweep generator. A voltage-controlled oscillator (VCO) is driven by the sweep voltage to produce a frequency sweep. The output of the VCO is amplified and passes through a variable attenuator. Although this conceptual block diagram is helpful in understanding the instrument operation, the actual implementation is much more complex. Also, as mentioned earlier, most function generators will take advantage of their DDS architecture and dynamically alter the phase increment register to cause the generator to sweep.

When using the sweep operation, the user can specify the start frequency, the stop frequency, the sweep time, and the sweep type (linear or logarithmic). The sweep may be set to repeat continuously or the sweep may be triggered via an external trigger input.

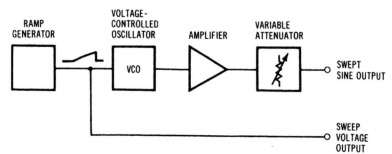

Figure 3.18   A conceptual block diagram of a sweep generator. A voltage-controlled oscillator is driven by a ramp voltage.

3.5.7   Sync Output

A *sync output* on a function generator provides an easy and reliable means of triggering other instruments or devices synchronously with the function generator waveform. Simply put, the sync signal provides a continuous output that represents the "pulse" of the waveform. Usually, the sync signal is a standard TTL logic level (or a reasonable size square wave). The sync signal provides for easy oscilloscope triggering even when the function generator's output is a complex waveform or is varying in amplitude.

When the function generator is producing a sine wave, square wave, or triangle wave, the sync signal is high when the output waveform is high and low when the output waveform is low. For arbitrary waveforms, the sync output will trigger once per cycle of the arbitrary waveform. For AM and FM, the sync output will track the *modulating* signal and not the carrier, to support easy viewing of the modulation characteristics on an oscilloscope.

3.5.8   Phase Locking

In a traditional signal source, the waveform is produced by one or more free-running oscillators, which are designed to operate over some frequency range. In a synthesized signal source (including DDS sources), one or more crystal-controlled reference oscillators operate at a fixed frequency, which is then used by the synthesizer circuitry to produce the desired output frequency. Because the reference oscillator does not have to change frequency, it can be made extremely stable. The synthesizer consists of digital circuits that have no frequency variation associated with them. So the source frequency is as stable and accurate as the reference oscillator. The result is a variable frequency source that has the precise frequency control associated with a fixed-frequency oscillator.

The reference oscillator is usually designed such that it can be electronically locked to other synthesized sources. Each source provides a *reference output* (usually 10 MHz) as well as a *reference input*. The reference output of one source is connected to the reference input of one or more other sources (Figure 3.19). These sources lock their internal reference oscillators to the reference output of the original source. This keeps all of the sources locked precisely in frequency to one reference oscillator. Two synthesized sources that are locked together exhibit essentially zero frequency drift between them. (Two free-running sources will drift apart in frequency quite noticeably.)

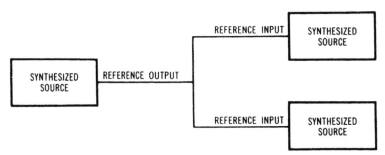

Figure 3.19 Multiple synthesized sources can be locked together in frequency using the reference inputs and reference outputs.

3.6 Pulse Generators

Pulse generators are specifically designed to produce high-quality square waves and pulses. They generally operate over a frequency range as low as 1 Hz and as high as 1 GHz. As shown in Chapter 1, a pulse train has a large number of significant harmonics requiring an instrument bandwidth much greater than the fundamental frequency. Above about 50 MHz, pulse generators are usually optimized for the difficult task of producing clean pulses. The bandwidth required to generate these pulses extends to hundreds of MHz, so high-performance pulse generators tend to have a correspondingly high price. Some function generators offer pulse capability to produce a combined pulse/function generator. For example, Figure 3.20 shows an 80-MHz function generator with pulse capability.

Figure 3.20 This function generator also includes arbitrary waveform and pulse capability. (Photo: Agilent Technologies. Reprinted with permission.)

For convenience, pulse generators are generally specified in the time domain. Waveform characteristics such as period, duty cycle, pulse width, pulse period, and frequency can be selected from the front panel. Distortion is not specified, since it is a sine wave specification that does not apply to pulses. Instead, pulse-related parameters like rise time and fall time are included. Usually very flexible control is included to allow positive-going pulses, negative-going pulses, and symmetrical waveforms (Figure 3.21).

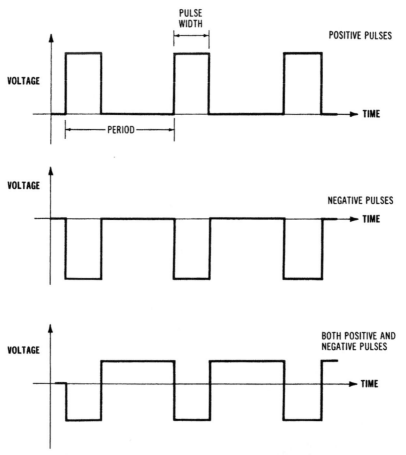

Figure 3.21   Flexible control over the pulse width, period, and polarity of the output pulses is provided on a pulse generator.

Pulse generators usually supply a triggering capability, which produces an output pulse in response to a trigger signal. Figure 3.22 shows how a trigger signal initiates an output pulse with a selectable amount of delay in between. This feature can be used to simulate a particular timing condition in a digital logic circuit. The trigger condition may also come from the push of a front panel button. This is handy for single stepping a sequential logic circuit or microprocessor. Many pulse generators provide burst modes which allow a fixed number of pulses to be output repetitively or in response to a trigger.

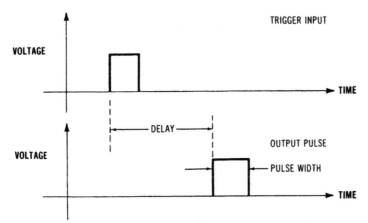

Figure 3.22 The trigger input causes a pulse at the generator's output after a controllable amount of delay.

Pulse generators are generally used in digital systems to simulate or replace digital signals such as clocks, data lines, and control lines. The pulse generator can also be used in analog systems. For example, square waves are used to test audio amplifiers and for making time domain reflectometer measurements (Chapter 5).

Entry-level pulse generators have a single output but some products have a second, optional channel available. Pulse generators are often used with complex digital systems, so having multiple output channels is a desirable feature.

3.7 RF Signal Generators

An *RF signal generator* is primarily designed to be used as a test stimulus for radio receivers, but also ends up being used as a general-purpose sine wave source in the radio frequency range. Signal generators are provided with flexible modulation capability, often with a built-in modulation source. Amplitude modulation and frequency modulation are most common, with phase and pulse modulation also sometimes provided (Figure 3.23). Some signal generators support the latest digital communications formats for testing of wireless communications equipment. Precision, wide-range attenuation, and low RF leakage are usually required in signal generators to support full-range receiver tests. Figure 3.24 shows a typical signal generator.

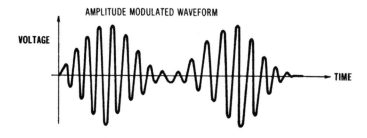

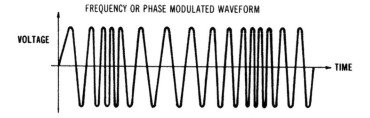

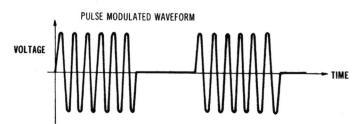

Figure 3.23   Typical waveforms produced by signal generators.

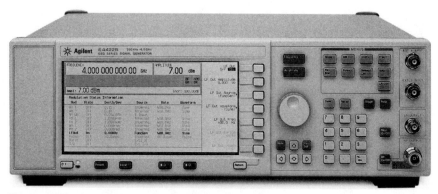

Figure 3.24   A radio frequency signal generator with an internal modulation source. (Photo: Agilent Technologies. Reprinted with permission.)

Figure 3.25 is the conceptual block diagram of a signal generator having an internal audio oscillator for use as a source of modulation. The heart of the signal generator is a voltage-controlled oscillator (VCO). The frequency of the VCO is determined by the voltage present at the control input. If the control voltage is increased or decreased, then the frequency of the VCO increases or decreases, respectively. In the FM position, the frequency of the VCO tracks the audio oscillator, resulting in a frequency-modulated signal. For amplitude modulation, a modulator circuit is added after the oscillator. The modulator circuit varies the amplitude of the VCO's output without modifying its frequency, producing an amplitude-modulated signal. The actual block diagram used to implement the signal generator is much more complicated than this, but the conceptual block diagram is a valid tool for understanding these signal sources.

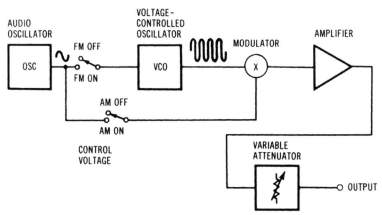

Figure 3.25 A conceptual block diagram of a signal generator with an audio source for modulating the carrier.

The frequency accuracy and stability are very important in a signal generator used to test radio receivers. The passband of the receiver is usually very narrow, much less than a percent of the center frequency. An even better frequency accuracy is required for the test signal. Harmonic distortion is not usually very critical in receiver testing since the harmonics are well away from the frequency of the receiver. (For other high-frequency applications, however, the harmonic distortion may be important.) Amplitude accuracy is critical for accurately measuring the receiver's sensitivity.

Higher-quality signal generators have extremely good close-in sideband performance (usually specified as residual FM or phase noise). This is important in radio and other systems where even very low levels of noise close to the carrier can degrade the system performance. For instance, the rejection of a signal that is very close to but outside a radio receiver's bandwidth is an important test of a receiver's performance. This adjacent channel rejection test can be performed by tuning the signal generator to a frequency very close to the receiver's frequency and then increasing the generator's output level until the receiver can no longer reject the signal. The signal generator must not have any close-in sidebands, otherwise these sidebands will spill over into the receiver passband and be detected as if they were a valid signal.

Example 3.2

Of the types of sources described in this chapter, which ones would suit the following applications: 1) Generate a 1-MHz digital clock. 2) Provide a 2-kHz sine wave for audio testing. 3) Generate a 146.52-MHz frequency modulated sine wave for testing a radio receiver?

1. Any source capable of outfitting a square wave, with perhaps some DC offset: function generator or pulse generator.

2. All of the sources except the pulse generator can output a sine wave, but signal generators usually don't go that low in frequency: sine wave source or function generator.

3. Although modulation is sometimes supplied in other sources, they typically don't go high enough in frequency: RF signal generator.

3.8   Summary of Signal Sources

A wide variety of signal sources are available on the market, many with overlapping capability. Table 3.1 summarizes the different categories covered in this chapter, highlighting the key differences. Because of the overlap often found between the categories of sources, this table is only a general guide as to the specifications and capabilities of each type of source.

Table 3.1   Table of Signal Source Characteristics

Signal Source	Frequency Range	Types of Waveforms	Key Features	Key Applications
Sine wave	1 Hz to 1 MHz	Sine	Spectral purity	Audio and low frequency test
Function generator	1 Hz to 15 MHz	Sine, square, triangle (may include arbitrary waveforms and pulse)	Often includes sweep, modulation, and burst	General-purpose electronic test

Table 3.1   Table of Signal Source Characteristics (Continued)

Signal Source	Frequency Range	Types of Waveforms	Key Features	Key Applications
Pulse generator	1 Hz to 1 GHz	Pulse	Precise pulses, including variable rise and fall times	Digital device test
RF signal generator	100 kHz to 2 GHz	Sine	AM/FM, may include digital modulation formats	Radio receiver testing

This table reflects the specifications of a typical instrument in each category. Actual specifications and capability will vary from instrument to instrument.

3.9   References

"Agilent 33120A Function Generator User's Guide," Agilent Technologies, Publication No. 33120-90005, August 1997.

"Agilent ESG Family of RF Signal Generators—Data Sheet," Agilent Technologies, Publication No. 5965-3096E, May 2001.

Coombs, Clyde F., Jr., *Electronic Instrument Handbook*, 3rd ed., New York: McGraw-Hill, 1999.

Helfrick, Albert D., and William D. Cooper. *Modern Electronic Instrumentation and Measurement Technique*, Englewood Cliffs, NJ: Prentice-Hall, 1990.

Jones, Larry D., and A. Foster Chin. *Electronic Instruments and Measurements*, 2nd ed., Englewood Cliffs, NJ: Prentice Hall, 1991.

Oliver, Bernard M., and John M. Cage. *Electronic Measurements and Instrumentation*. New York: McGraw-Hill, 1971.

"Signal Generator Spectral Purity," Hewlett-Packard Company, Application Note 388, Publication No. 5952-2019, 1990.

CHAPTER 4

Oscilloscopes

The oscilloscope is probably the most versatile electronic test instrument in use today. Unlike meters, which only allow the user to measure amplitude information, the oscilloscope (or "scope") allows the user to view the instantaneous voltage versus time. Not only does this reveal details of the waveform to the scope user, it also allows the measurement of parameters such as frequency, period, phase, rise time, and overshoot.

There are two main classes of oscilloscopes: analog oscilloscopes and digital oscilloscopes. Early scopes were analog in nature and for many years analog scopes were the workhorse of the electronics industry. More recently, digital technology has largely overtaken analog, and the majority of the oscilloscopes sold are digital oscilloscopes (or *digitizing oscilloscopes*). Analog scopes are still commonly used and are preferred by some users for certain applications.

4.1   The Oscilloscope Concept

The primary function of an oscilloscope is to display an exact replica of a voltage waveform as a function of time. This picture of the waveform can be used to determine quantitative information such as the amplitude and frequency of the waveform as well as qualitative information about the waveform shape. The oscilloscope can also be used to compare two different waveforms and measure their time and phase relationships.

Most oscilloscopes have two input channels. Single-channel scopes are sometimes available, but have been largely eclipsed by two-channel and four-channel models. Four-channel scopes are particularly useful for measuring signals in digital systems, which usually have many signals that must be viewed simultaneously. Four-channel scopes operate similar to two-channel scopes, with the exception of having twice as many inputs and the ability to display four waveforms simultaneously. As mentioned earlier, digital scopes have largely replaced analog scopes for most applications, so the emphasis in this chapter is on the digital instrument. In concept, analog oscilloscopes and digital oscilloscopes perform the same types of measurements, but with different techniques internal to the instrument. These differences will be addressed later in the chapter. An example of a four-channel oscilloscope is shown in Figure 4.1.

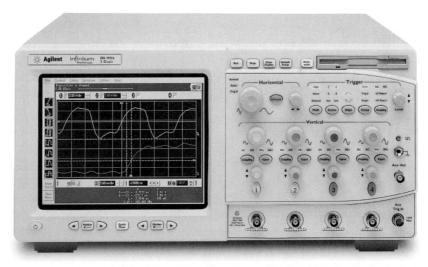

Figure 4.1   A four-channel 500-MHz digital oscilloscope. (Photo: Agilent Technologies. Reprinted with permission.)

A sine wave is displayed by an oscilloscope, as shown in Figure 4.2. The vertical axis represents voltage and the horizontal axis represents time. This results in the same voltage versus time plot for a sine wave, which was introduced in Chapter 1. The oscilloscope display has a set of horizontal and vertical lines called the *graticule,* which aid the user in estimating the value of the waveform at any particular point. (The graticule is shown in Figure 4.2, but is omitted in other figures.) Fundamentally, the concept of an oscilloscope is simple: accurately produce the voltage versus time plot of a waveform. However, actually accomplishing this may be more complicated.

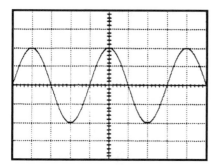

Figure 4.2 The oscilloscope display with a sine wave connected to the scope input.

4.2 Analog Scope Block Diagram

A simplified block diagram of an analog oscilloscope is shown in Figure 4.3. The display of the oscilloscope (a *cathode ray tube* or *CRT*) has two inputs to it, the vertical position to be plotted and the horizontal position to be plotted. These inputs connect to the deflection plates of the CRT and control the position of the beam that plots the waveform on the display. The vertical axis represents the waveform to be plotted, so the input of the oscilloscope is connected to an amplifier, which drives the vertical input. The horizontal axis is time, so the horizontal input is driven by the *timebase generator*, which generates a signal that represents increasing time. We know what the vertical signal looks like—it's just whatever the input waveform is, perhaps a sine wave or square wave. What the timebase needs to do is constantly increase the horizontal position as the input voltage goes through a cycle. This results in the input waveform being swept or painted across the display at some rate, depending on the timebase.

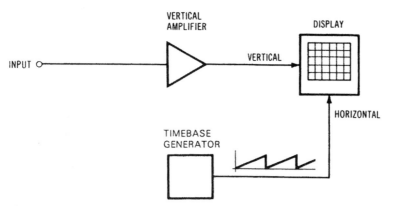

Figure 4.3 Conceptual block diagram of a basic analog oscilloscope.

The timebase waveform required to do this is a constantly increasing voltage called a *ramp waveform*. Since the display is usually updated repetitively, the ramp starts over when it reaches its maximum value, resulting in a *sawtooth waveform* (Figure 4.4). Each cycle of the sawtooth waveform corresponds to one sweep across the display. A small delay (called the *retrace time*) occurs between each ramp of the sawtooth as the horizontal position is reset to the left side of the display.

The sawtooth waveform of the timebase is started in response to a trigger event, which corresponds to a specific location on the waveform. The trigger initiates the sweep across the display and ensures that the waveform is kept stable from sweep to sweep.

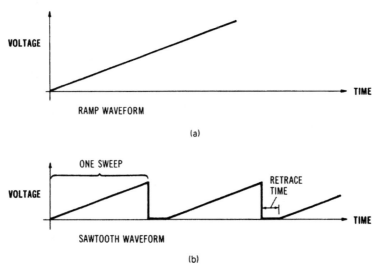

Figure 4.4   a) The timebase generator produces a ramp voltage. b) Repeating multiple ramps result in a sawtooth waveform.

4.3   Digital Scope Block Diagram

Conceptually, analog and digital oscilloscopes do the same thing—they display voltage waveforms. While the analog scope uses traditional circuit techniques to display the voltage waveform on a CRT, a digital scope converts the analog input signal to digital form as a series of binary numbers that are used to produce the waveform on the display. This means that a digital scope is inherently a storage scope, because the waveform is stored digitally. (Digital scopes are often called *digital storage oscilloscopes,* or DSOs.)

The conceptual block diagram of a digital scope is shown in Figure 4.5. The analog input is increased by the amplifier, which drives the *analog-to-digital converter* (ADC). The ADC converts the analog signal into digital form at each pulse of the sample clock. The resulting digitized waveform is stored in the *memory* as a series of digital numbers. The *dis-*

play processor converts the stored waveform data into the format required for the display. While the display could be an analog CRT, it is more likely to be a pixel-based, digital display such as a flat-panel *liquid crystal display* (LCD).

The acquisition of the signal in digital form is in response to the trigger event, just as the analog scope sweep is started by the trigger. However, the digital scope may be configured to start acquiring the waveform ahead of the trigger event, which allows the user to look ahead of the trigger. Still, the basic principle holds that the acquisition of the signal is synchronized by the trigger event.

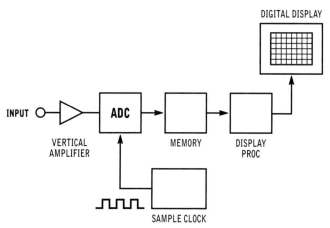

Figure 4.5 Conceptual block diagram of a digital oscilloscope.

4.4 Sample Rate

Figure 4.6 shows the result of this sampling process in the digital scope. The waveform is no longer a continuous analog voltage but has been sampled at a constant sampling period, according to the sample clock operating at a frequency of f_S.

$$f_s = 1/T_S$$

where

f_s = sample rate

T_S = sample period

The sampling process introduces gaps in the waveform but as long as the waveform is sampled fast enough, these gaps can be ignored, or at least accommodated.

Capturing the waveform in digital form provides some great benefits. The waveform can be stored for later viewing or transferred to a computer to become part of a written report. The math functions can be used to add or subtract two waveforms or more advanced signal processing can extract additional information from the waveform.

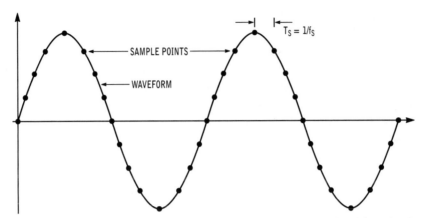

Figure 4.6 The analog input is sampled and converted into digital form by the digital oscilloscope.

The *sampling theorem* states that for a baseband signal to be accurately reproduced in sampled form, the sample rate must be greater than twice the highest frequency present in the signal.

$$f_s > 2\,BW$$

where

f_s = sample rate
BW = bandwidth of the signal

The minimum required sample rate is known as the *Nyquist rate*. This relationship guarantees that slightly more than two samples per cycle are acquired on a sine wave having a frequency right at the bandwidth of the signal. This equation defines the minimum theoretical sample rate. In practice, a higher sample rate is required to accurately represent the signal.

Example 4.1

What is the minimum sample rate required by the Sampling Theorem to completely represent a signal having a 100-MHz bandwidth?

The Sampling Theorem requires

$f_s > 2\,BW$

$f_s > 2 \times 100$ MHz = 200 MSa/sec

The sample rate, f_s, must be greater than 200 MSa/sec.

4.4.1   Aliasing

If the sample rate does not meet the requirements of the Sampling Theorem, *aliasing* can occur. With an insufficient sample rate, one frequency can show up as an *alias* of another frequency.

Figure 4.7 shows two waveforms that share the same set of sample points. The lower frequency signal with one period shown in the figure is sampled fast enough to prevent aliasing. The higher frequency signal with nine periods shown in the figure is not being sampled fast enough to be accurately captured. At this sample rate, the higher frequency signal is indistinguishable from the lower frequency signal, since they share the same sample points.

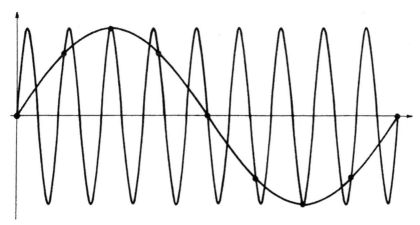

Figure 4.7 These two waveforms share the same set of sample points. The lower frequency signal (one period shown) will be accurately captured with the sample rate shown. The sample rate is too slow for the higher frequency signal.

4.4.2   Sample Rate Specification

The sample rate and bandwidth specifications on a digital scope tend to be confused. Does a 100-MHz digital scope have a sample rate of 100 MHz or a bandwidth of 100 MHz? Or maybe both? To help alleviate this problem, digital scope manufacturers have standardized on specifying the sample rate in samples per second and the bandwidth in Hertz. More typically, the sample rate is specified in gigasamples per second (GSa/sec) or megasamples per second (MSa/sec) and the bandwidth is given in Gigahertz or Megahertz.

This still leaves the user with the question of how does sample rate and bandwidth relate in a practical digital scope? Or how much sample rate is required for a particular bandwidth? The answer depends on the sampling technique used in the oscilloscope.

4.5   Real-Time Sampling

Real-time sampling is the most obvious and intuitive type of sampling. This type of sampling simply acquires sample points uniformly spaced in time and fast enough to capture the entire bandwidth of the signal (see Figure 4.8). All of the sample points are

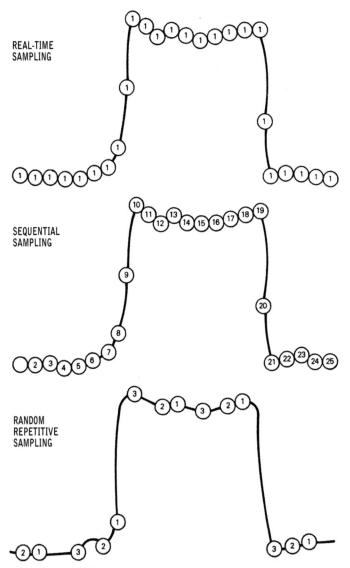

Figure 4.8   Digital scope sampling techniques. The numbers on the samples indicate which acquisition the sample is collected on. Real-time sampling captures all of the samples on the first acquisition, while sequential and random-repetitive sampling require multiple acquisitions to capture all of the samples required. (Agilent Technologies. Reprinted with permission.)

acquired in one acquisition of the waveform (one measurement or one trigger event). The major advantage of this technique is that a one-time transient (or *single-shot event*) can be acquired. The main disadvantage is that the analog-to-digital converter (ADC) must operate well above the Nyquist rate in order to correctly capture the signal, which may increase the cost of the oscilloscope.

Real-time sampling can provide waveform information in front of the trigger event. This pretrigger information is obtained by letting the sampling system continuously acquire data, which is retained and displayed when a trigger event finally occurs. Pretrigger viewing is a powerful advantage of the digitizing scope architecture, since the user can trigger on some failure condition and then look back in time to determine what caused the failure. (The time before the trigger event is sometimes referred to as *negative time*.)

Most oscilloscopes that use real-time sampling use interpolation between sample points to reconstruct the waveform when operating near the Nyquist rate. This interpolation fills in the gaps between the sample points and provides a good estimate of the waveform voltage between samples. This interpolation algorithm usually takes the form of a digital filter following the ADC (Figure 4.9).

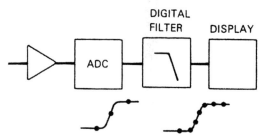

Figure 4.9 A digital reconstruction filter provides interpolation between sample points when using real-time sampling.

The single-shot bandwidth of a digital scope is defined as the maximum bandwidth that can be captured in a single-shot measurement. The relationship between the single-shot bandwidth of a real-time scope and the sample rate is given by

$$BW_S = f_s/k_R$$

where BW_S is the single-shot bandwidth

f_s = the sample rate

k_R = a factor depending on the reconstruction technique

The factor k_R can theoretically approach 2 (according to the Sampling Theorem), but is typically 2.5 or greater. With $k_R = 2.5$, reconstruction is suitable only for sine wave signals, since the reconstruction filter shape must be steep, which causes ringing on pulsed signals. For a typical transient phenomenon such as a pulse, a factor of $k_R = 4$ allows a reconstruction filter with good transient response to be used. With $k_R = 4$, a typical oscilloscope having a sample rate of 1 GSa/sec has a single-shot bandwidth of 250 MHz. In cases where no reconstruction is used, $k_R = 10$ is required, giving 10 samples per period of the highest frequency sine wave.

Example 4.2

An oscilloscope uses real-time sampling with digital reconstruction using $k_R = 4$. If the sample rate of the scope is 400 MSa/sec, what is the single-shot bandwidth of the scope? If no reconstruction is used, what is the single-shot bandwidth?

The single-shot bandwidth is given by

$$BW_S = f_s/k_R$$

For the $k_R = 4$ case,

$$BW_S = (400 \times 10^6)/4 = 100 \text{ MHz}$$

With no reconstruction used, $k_R = 10$, therefore,

$$BW_S = (400 \times 10^6)/10 = 40 \text{ MHz}$$

4.6   Repetitive Sampling

Repetitive sampling (also called *equivalent-time sampling*) includes two different sampling techniques: *sequential sampling* and *random-repetitive sampling*. These two sample techniques are shown in Figure 4.8. Either sampling technique can be used if the following two conditions are met.

1. The waveform must be repetitive.
2. A stable trigger event must be available.[1]

Since the waveform is repetitive, the scope can acquire the waveform samples over many periods of the waveform. Repetitive sampling relies on the oscilloscope getting multiple opportunities to "see the waveform go by." A stable trigger is required so that the scope can arrange these samples captured from different acquisitions properly on the display.

---

1.   Most oscilloscope measurements require a stable trigger. This requirement is mentioned here to emphasize how repetitive sampling uses the trigger event to keep track of where the samples occur.

Sequential sampling operates by acquiring a single sample on each trigger or acquisition (see Figure 4.8). When a trigger occurs, a sample is taken at a precisely controlled time after the trigger. On each subsequent trigger, this delay time is increased so that all portions of the waveform are eventually sampled. As long as the time delay from the trigger point to the acquired sample is precisely controlled, the waveform is accurately reproduced. The chief advantage of this technique is that a much slower ADC can be used. For example, this technique has been used to implement microwave scopes with bandwidths of 50 GHz using an ADC with a sample rate of only 10 kSa/sec. Real-time sampling of this bandwidth would require an ADC sample rate of well over 100 GSa/sec, which is not cost effective.

Because all of the samples occur after the trigger event, sequential sampling provides no pretrigger information. Random-repetitive sampling overcomes the pretrigger limitation of sequential sampling. By sampling randomly (even before a trigger event occurs), some of the acquired samples end up occurring in front of the trigger. The timebase circuit of the scope must measure the time between the trigger event and the digitized samples, so that the samples can be placed properly in time. As shown in Figure 4.8, random-repetitive sampling may acquire multiple samples per trigger event, depending on the scope's timebase setting and the sample rate of the scope.

Since repetitive sampling techniques require multiple "looks" at the waveform, we might be concerned that this would cause the scope to respond slowly to waveform changes. For most common waveforms, the trigger rate of the signal is so fast that this is not usually a problem. Even though the waveforms are updated from multiple acquisitions, this can happen so fast that the scope's responsiveness is quite good. However, if the trigger rate is slow enough, the "building up" of the waveform can be objectionable. The most important issue occurs when there is one opportunity to capture the signal—a single-shot event—where repetitive sampling is useless.

Since random-repetitive sampling provides pretrigger information, it has largely displaced sequential sampling, except at microwave frequencies. At microwave frequencies, the timebase setting on the scope can be very small, causing the window of time that is viewed on the display to also be very small (perhaps 100 ps). The probability of a randomly acquired sample falling into the desired time window is so small that random-repetitive sampling would take a long time to acquire the entire waveform. Alternatively, sequential sampling forces the sample points to occur within the desired time window so the entire waveform can be acquired quickly.

Many digital scopes combine both real-time and repetitive sampling techniques, using real-time sampling for lower frequency bandwidths and repetitive sampling on the higher frequencies. Some scopes make the switch between real-time and repetitive automatically according to the time/division setting chosen. Other scopes provide a front panel control allowing the scope user to choose the type of sampling.

4.6.1   Effective Sample Rate

It may appear that repetitive sampling violates the Sampling Theorem since a slow ADC is being used to sample a high-frequency signal. The concept of effective sample rate is introduced to explain how this works.

$$f_{eff} = 1/T_{eff}$$

where

f_{eff} = the effective sample rate

T_{eff} = the effective sample period, which is equal to the time between samples after the waveform is acquired.

T_{eff} is not dependent on the sample rate of the ADC, but is dependent on how precisely the samples can be placed in time relative to the trigger event. After the entire waveform is acquired, T_{eff} is the time between adjacent sample points. Effective sample rate is really a measure of the timing resolution of the instrument. In a repetitive scope, the effective sample rate will be much higher than the actual ADC sample rate and will satisfy the Sampling Theorem. For example, a repetitive sampling scope might have a 20 MSa/sec sample rate but T_{eff} might be 100 ps. With this timing resolution, f_{eff} = 10 GSa/sec.

4.6.2   Memory Depth

The memory depth is important in real-time sampling, since it is often desirable to acquire long time records at a high sample rate. For a given sample rate, deeper memory allows longer time records to be acquired. Of course, the sample rate can also be decreased, allowing longer events to be digitized, but at the expense of lower sample rate and poorer timing resolution. Memory depth is particularly important for measuring single-shot events, especially if the exact event is long in duration.

The larger the memory size (memory depth or time record length), the more time the scope can sample with a given sample rate. The total time acquired by the scope acquisition system, T_{ACQ}, is given by

$$T_{ACQ} = N_S T_S = N_S/f_S$$

where

N_S = number of samples in memory

T_S = sample interval

f_S = sample rate

Memory depth is easy to overlook as a key scope specification, so a numerical example can help illustrate the tradeoff between memory depth and sample rate. Suppose a scope has a maximum sample rate of 1 GSa/sec, a 2000-point memory depth, and timebase settings from 5 seconds per division to 2 nsec per division. The maximum time that can be acquired by the scope at its maximum sample rate is:

$$T_{ACQ} = N_S/f_S = 2000 / (200 \times 10^6) = 10 \text{ μsec}$$

Assuming 10 divisions across the display, an acquisition time of 10 μsec corresponds to 1 μsec per division. For timebase settings slower than 1 μsec per division, the maximum sample rate cannot be maintained. In the extreme case of 5 seconds per division, the resulting sample rate is given by:

$$f_S = N_S/T_{ACQ} = 2000 / (5 \times 10) = 40 \text{ samples per second}$$

Clearly, memory depth can have a dramatic effect on the actual sample rate that can be delivered at slow timebase settings. Peak detect, discussed later in this chapter, is an important acquisition technique that can compensate for limited memory depth.

4.7 Triggering

The timebase performs an important function in the scope, but the timebase needs a trigger event to know when to initiate a sweep or waveform acquisition. Without this information, the waveform will be displayed with a random start time. At best, this results in an unstable waveform wandering across the display. At worst, the display is a jumbled collection of waveforms smeared together, filling the entire display area (Figure 4.10a).

A properly triggered waveform, having a predictable and repeatable starting point, will be stationary from sweep to sweep (Figure 4.10b). Consider the sine wave shown in Figure 4.11a. If the trigger circuit is set up to trigger at the start of the sine wave, then triggers will occur at the points identified in the figure. Each piece of the sine wave is then displayed, with the trigger point at the leftmost edge of the display. Notice that each trigger point occurs at the same place on each cycle. This will result in a stable display, as shown in Figure 4.11b. The figure implies that none of the cycles of the waveform are missed by the oscilloscope. In reality, since the oscilloscope may not be able to prepare for a new sweep instantaneously (due to retrace time), some cycles may be missed.

Triggering is probably the most troublesome part of using an oscilloscope. Most scopes provide a variety of ways to trigger on a signal so that the user can customize the triggering to a particular measurement problem. Of course, this also means that the user must understand and make decisions about what type of triggering to use.

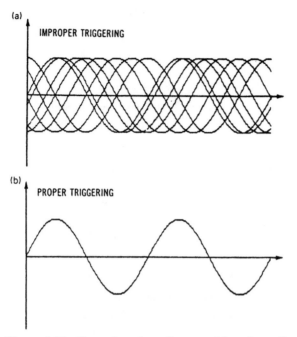

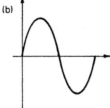

Figure 4.10 Examples of oscilloscope triggering with a sine wave input. a) Improperly triggered waveform results in an unstable display. b) Properly triggered waveform has a stable display.

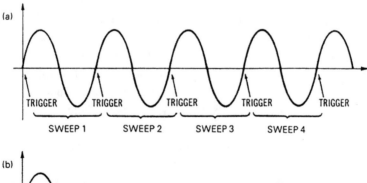

Figure 4.11 a) The sine wave with multiple cycles and multiple trigger points. b) The resulting display when the scope is properly triggered.

4.7.1   Edge Trigger

Edge trigger is the simplest form of oscilloscope triggering. The trigger point on a waveform is defined using a voltage level and a slope (positive or negative). The *trigger level* control determines the voltage level at which the trigger will occur. The *trigger slope* control determines whether the trigger will occur on a *rising edge* (positive slope) or *falling edge* (negative slope) of a waveform. Figure 4.12a shows a waveform and some possible trigger times. All of the trigger points shown are at a trigger level of 0.5 volts, but some are for a positive slope and the others are for a negative slope. The resulting scope displays for the two triggering conditions are shown in Figures 4.12b and 4.12c.

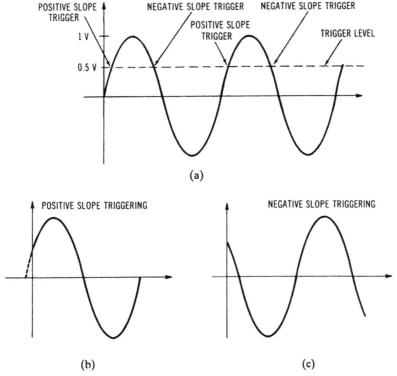

Figure 4.12   a) Waveform shown with positive slope and negative slope trigger points. b) The scope display with positive slope triggering. c) The scope display with negative slope triggering.

4.7.2   Trigger Source

When the scope is triggering on a waveform on one of the inputs, the oscilloscope is using an *internal trigger*. Multichannel scopes allow the user to select any of the channels for use as the trigger source. An *external trigger* input is provided to allow the user to connect an external signal to the oscilloscope to be used for triggering purposes. This signal usually cannot be viewed on the scope display, but some scopes have a *trigger view* feature to aid in setting up the triggering. The *line trigger* selection uses the AC power line voltage as the trigger signal (usually 60 Hz or 50 Hz, depending on the regional power system standards). Line trigger is useful for observing waveforms that are either directly related to the power line frequency (including its harmonics) or have power line-related voltages superimposed on the original waveform.

The trigger may be AC coupled or DC coupled. DC coupling presents the trigger circuit with a waveform containing both AC and DC voltages, while AC coupling removes any DC that is present. AC coupling is useful when the desired trigger signal is riding on top of a DC voltage that needs to be ignored. Most oscilloscopes include additional filters that can be switched into the trigger circuit to condition the trigger signal. The *low frequency reject* filter will remove low frequencies such as the 60-Hz line frequency (and its harmonics), which may be present in the trigger signal, causing triggering problems. The *high frequency reject* filter will similarly remove high-frequency noise, which also may cause triggering problems.

Noise reject rejects, reduces, or eliminates false triggering on noisy waveforms. Unlike low-frequency reject and high-frequency reject, noise reject works without changing the frequency response of the trigger system.

4.7.3   Trigger Holdoff

The *trigger holdoff* control disables the triggering circuit for a period of time after a trigger occurs. This is useful when the waveform has several places in its cycle that are the same as the trigger condition. The oscilloscope would normally trigger on all of them, but with the trigger holdoff, the scope ignores all but the first one. Suppose we want to display the first cycle of the pulsed sine wave shown in Figure 4.13. The trigger could be set to zero volts and positive slope, which will trigger at the beginning of the sine wave. Unfortunately, this trigger condition exists at the start of every one of the sine wave periods, so the oscilloscope will display every cycle. However, if the trigger holdoff is adjusted properly, the subsequent triggers can be ignored and only the first cycle of each sine pulse will be displayed.

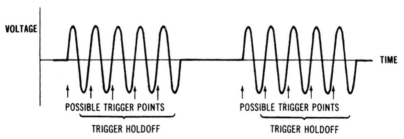

Figure 4.13   Example of trigger holdoff operation. The trigger holdoff disables the oscilloscope's trigger circuit for an adjustable length of time.

While setting the trigger holdoff as a fixed time, called *time holdoff*, is most common, some scopes provide an additional holdoff feature, called *event holdoff*. Instead of specifying the amount of holdoff as a fixed time, the user specifies the number of trigger events that need to be ignored. This is useful in cases where a fixed number of pulses needs to be ignored and the total time associated with these pulses varies significantly (such as synchronization pulses from a disk drive or other rotating machine that is varying in speed).

4.7.4   Video Trigger

Since composite television and video signals can be quite complex, they don't lend themselves to simple edge triggering. Therefore, many scopes provide a TV or *video trigger* capability, which extracts the horizontal and vertical sync signals from the input video and uses them for triggering the scope. The simplest form of video triggering lets the user trigger on all horizontal sync pulses (all lines) or all vertical sync pulses (all fields). More advanced triggering systems let the user pick the individual line and field to trigger on, allowing any particular portion of the video waveform to be easily evaluated.

4.7.5   Pulse-Width Trigger

Pulse-width trigger, also called *glitch trigger,* monitors the width of the pulses on a scope channel and triggers when they meet a specified condition. The situation shown in Figure 4.14 has the trigger condition set for pulses that are less than 100 nsec wide. Pulses wider than 100 nsec are ignored while a narrower pulse causes the scope to trigger. Pulse-width triggers can also be set to trigger when the pulse width is greater than a specified time or when the pulse width is within a range of values (specified as a maximum and minimum pulse width).

Digital glitches can be the source of significant and infrequent failures in digital circuits, making them very difficult to debug. Glitch trigger allows a designer to find narrow pulses before they cause problems.

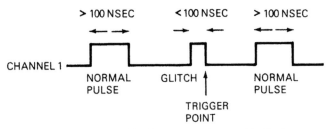

Figure 4.14   Pulse-width triggers can be used to find narrow glitches in digital circuits.

4.7.6   Pattern Trigger

Single-channel triggering is sufficient for many applications, but the multitude of signals found in complex digital systems requires a more sophisticated triggering system. *Pattern trigger* expands on the usual level and slope triggering by letting the user define the trigger condition in more powerful terms. Each of the scope channels is considered as a logic level, with a logic high (H) being any voltage greater than the trigger level and a logic low (L) being any voltage below the trigger level. Any scope input can be ignored for triggering purposes by specifying its trigger condition as a "don't care" (X).

The pattern trigger feature lets the user define a trigger condition by specifying a logic pattern that must be present on the scope's inputs. As soon as this pattern appears, the scope triggers. Figure 4.15 shows the input waveforms on the four input channels of a scope. With the trigger level set near the middle of the logic swing, the scope is set to pattern trigger on the pattern H H L L (High, High, Low, Low) on channels 1 through 4. The scope triggers at the point shown in Figure 4.15.

This feature can be used to trigger on such things as a particular digital value on an address or data bus. Suppose that in a microprocessor system, writing to a peripheral device is causing a system failure. Pattern trigger could be used to trigger on the appropriate address and enable lines associated with writing to the device. Once the trigger condition is established, the scope can be used to view other signals in the system to determine the cause of the problem.[2]

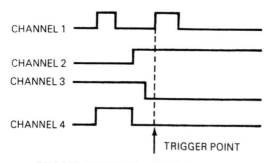

Figure 4.15   Example of pattern trigger operation.

A more powerful version of pattern trigger is *time-qualified pattern trigger*. In addition to defining a logic pattern, the user specifies how long the pattern must be present for a valid trigger to occur. The time limit can be specified as greater than a limit, less than a

---

2.   Pattern trigger is an especially useful feature when combined with the high channel count of the mixed-signal oscilloscope (see Section 4.19).

limit, or within some range. This provides the scope user with much more control of when a trigger occurs. In many digital circuits, it makes a big difference how long a particular logic pattern is present.

When used on a single channel, by setting the other channels to "don't care," time-qualified pattern trigger operates the same as pulse-width trigger. If the other channels are set high or low, they can be used to qualify the pulse-width trigger. Suppose narrow glitches are undesirable only when an enable line is high. Time-qualified pattern trigger can be set up to trigger on a narrow pulse, but only when the enable line is high, providing a more complete description of the desired trigger event.

4.7.7 State Trigger

State trigger configures the trigger system such that one channel of the scope is defined as the clock of the trigger system and the other channels are read on the rising or falling edge of the clock. For example, channel one might be configured as the clock (rising edge) and the trigger condition on the other channels might be L H L (Figure 4.16). On each rising edge on channel one, the trigger system evaluates the logic level of each channel and when the state of the channels matches the trigger condition, a valid trigger has occurred.

State triggering has the most utility in situations where logic levels are only valid on a clock edge. For example, a D type flip-flop changes its output based on the state of its D input when a rising edge appears on its clock line. The logic level of the D input is significant only when a rising clock edge occurs. Time-qualified pattern trigger could be used to trigger on some event such as a glitch at the D input, but would not be useful for triggering on a particular logic state *valid at the clock edge*. However, state trigger fires *only* at the clock edges and *only* when the specified logic state occurs simultaneously.

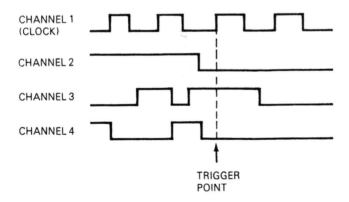

TRIGGER CONDITION = STATE ↑ L H L

Figure 4.16 State trigger triggers on a pattern, but only on clock edges.

4.7.8   Transition Time Trigger

Conventional scope triggering makes use of a single trigger level. *Transition time trigger* uses two trigger levels, producing a trigger based on the time the waveform spends between the two trigger levels (Figure 4.17). This capability allows the scope to trigger on a *rise time* or *fall time* condition. Normally, the trigger levels are set to the high and low logic thresholds or to the 10% and 90% points of the waveform. The scope can be set to trigger on transition times that are greater than or less than a specified time. For example, a digital designer can use this feature to trigger on excessively slow rise times that could cause problems in a digital design.

Figure 4.17   Transition time triggers monitor the time spent between two trigger levels, triggering when that time meets the specified criterion.

4.7.9   Setup and Hold Trigger

Digital circuits often use a clock signal to synchronize the transfer of logic signals from one device to another. Flip-flops, latches, and registers are common examples of digital circuits that use such a clock. For reliable circuit operation, the data must be stable logic high or low for a specified time called the *setup time* before the clock edge occurs. Similarly, the data must be stable high or low for a specified time called the *hold time* after the clock edge. Setup time and hold time are illustrated in Figure 4.18.

Setup time trigger triggers when the specified setup time is violated. That is, the data signal is not a stable valid high or low during the time window before the clock edge. Setup time trigger uses two scope channels, with one connected to the data signal and the other connected to the clock signal. *Hold time trigger* works very much the same way. A trigger occurs if the data is not stable for the prescribed time *after* the clock edge.

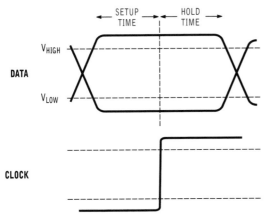

Figure 4.18 Setup time refers to the time required for the data to be a stable valid high or low before the clock edge. Hold time refers to the time required for the data to be stable after the clock edge.

4.8 Acquisition/Sweep Control

Scopes have a number of different modes that control how waveforms are acquired. In an analog scope, these are usually called "sweep" modes. Digital scopes don't "sweep," so other terms like "acquisition" modes are used. In this section, we use the term "acquisition."

In the *single* mode, the oscilloscope displays only one acquisition and then waits until the user resets the scope for another acquisition. This allows the user to capture a particular single-shot event, rather than continuously updating the display with unwanted information. Although this feature is supplied on analog oscilloscopes, it is most useful on digital scopes since the waveform can be stored in memory and maintained on the screen. On a conventional analog scope, the single sweep mode tends to be a brilliant, but brief, flash across the display.

4.8.1 Automatic and Normal Acquisition

Most oscilloscopes supply two types of continuous acquisitions: *automatic* and *normal*. The difference between the two is rather subtle, but important. With normal mode, the oscilloscope acquires a new waveform whenever a valid trigger occurs. This is fine for most AC waveforms, but is inconvenient if the input is a DC voltage. In this case, no trigger will occur since the DC input is constant and will not be passing through whatever trigger level happens to be selected. The result is a blank display.

Automatic mode operates just like normal mode except that if no trigger occurs for a period of time (as in the case of a DC input or an incorrectly adjusted trigger level) the scope acquires the waveform anyway. This is convenient because it gives the user a look at the waveform, even if it isn't triggered properly. Then, the user can take whatever action is necessary for correct triggering. In the case of a pure DC voltage, triggering is rather mean-ingless—the user just wants to view the constant voltage—so automatic mode works well. If a trigger starts occurring at a fast rate (typically anything greater than 40 Hz) then the scope triggers just like the normal mode. Automatic mode should be the default choice since it will give good results on most waveforms. The exception is when the desired trigger condition occurs so infrequently (a very low frequency sine wave, for example) that the automatic mode will begin triggering before the appropriate time. In that case, normal mode should be used.

4.8.2 Chop and Alternate Modes

Analog oscilloscopes normally have CRT displays that can only truly display one wave-form at a time. To handle more than one waveform simultaneously, most analog scopes elec-tronically switch between the two (or more) input signals and feed a composite signal to the display (Figure 4.19). If this is done quickly enough, it will not be noticeable to the user nor will it affect most measurements. (This consideration does not apply to digital scopes, which are not limited by the type of display, but by the analog-to-digital conversion circuitry.)

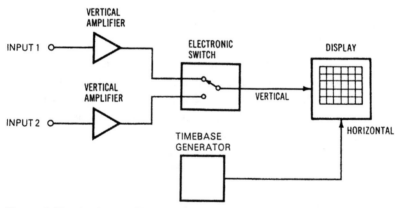

Figure 4.19 Analog oscilloscopes use an electronic switch to feed multiple channels to the display to show multiple waveforms simultaneously.

There are two different methods of controlling the electronic switch. *Chop mode* switches between the two inputs as fast as possible while the waveforms are being plotted on the display. Figure 4.20 shows how chop mode can be used to display two different waveforms simultaneously. For clarity, channel 1 is shown as a triangle wave and channel 2 is shown as a square wave. The display trace switches back and forth ("chops") between the two waveforms during the sweep, resulting in both of them appearing on the screen. Figure 4.20 exaggerates the chop effect to illustrate the point. In practice, the two channels are switched fast enough so that the chopping cannot be seen. This works well for sweeps that are much slower than the rate at which the two channels chop, so chop mode is best for low-frequency signals. If chop mode is used on high-frequency waveforms (and with very fast sweeps), then the chopping effect is noticeable on the display (as illustrated in Figure 4.20).

On the other hand, *alternate mode* takes a complete sweep of one waveform, then switches to the other input and takes another sweep. As shown in Figure 4.21, the first sweep displays channel 1, the second sweep displays channel 2, the third sweep displays channel 1 again, and so on. This works well during fast sweeps, because if the display update rate is fast enough, the user perceives that both inputs are being plotted simultaneously. Therefore, alternate mode works best with high-frequency signals. If alternate mode is used on low-frequency waveforms (and with slow sweeps), the individual sweeps of each channel become apparent and the effect of simultaneously displayed channels is lost. One potentially misleading problem with alternate mode is that even though the two waveforms appear to be displayed simultaneously, they really were measured at two distinct points in time (on two different triggers).

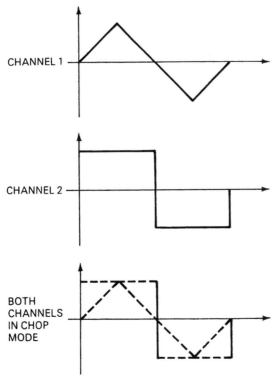

Figure 4.20   In chop mode, the display switches between the two channels at a fast rate.

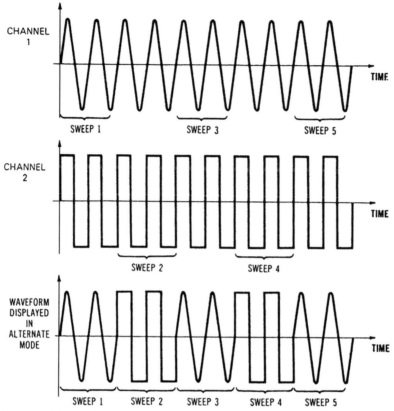

Figure 4.21 In alternate mode, complete sweeps from each channel are alternated one after the other, creating the effect of simultaneous waveforms.

4.9   Vertical Amplifier

The gain of the vertical amplifier (refer back to Figures 4.3 and 4.5) determines how large the waveform appears on the display. This allows the user to adjust the displayed signal so that accurate measurements can be made. The *vertical sensitivity* control determines the gain of the vertical amplifier and is calibrated in volts per division so that the user can determine the amplitude of the signal by noting the number of vertical divisions on the display.

Oscilloscopes have a vertical amplifier for each of their input channels. In the usual timebase mode of the oscilloscope, both channels are displayed as a function of time. In addition to displaying channel 1 and/or channel 2, most scopes provide the capability of displaying 1 + 2 or 1 – 2. Also, one or both of the channels may be capable of being displayed inverted (with its polarity reversed). (1 – 2 might not be provided, but 1 + 2 with channel 2 inverted can achieve the same result.) On an analog scope, these functions must be supplied in analog form in the vertical amplifier. Digital scopes often take advantage of their digital architecture and implement $1 + 2$ and $1 - 2$ as math functions that operate on the digital waveform.

4.10 Vertical Resolution

The ADC in a digital scope converts the input voltage to a binary number. This process results in *quantization* of the signal. That is, a continuously varying input waveform is converted into a series of discrete values.

A binary number having N bits can represent 2^N different values or codes. For a voltage range of V_{RANGE}, the voltage resolution of an N-bit converter is given by

$$V_{RES} = \frac{V_{RANGE}}{2^N}$$

Example 4.3

What is the voltage resolution of a digital oscilloscope with an 8-bit ADC when the sensitivity is 200 mv/div? (Assume the ADC range is spread over 8 vertical divisions.)

The voltage range covered by the ADC is 8×200 mV = 1.6 volts, so the resolution is given by

$$V_{RES} = \frac{1.6}{2^8} = 6.3 \, mV$$

The voltage resolution can also be expressed as a percent of the full-scale voltage range. Table 4.1 shows the vertical resolution for various numbers of ADC bits.

Table 4.1 Analog-to-Digital Converter Bits and Vertical Resolution

Number of Bits	Vertical Resolution
6	1.56%
7	0.78%
8	0.39%
9	0.20%
10	0.098%
11	0.049%
12	0.024%

4.11 AC and DC Coupling

Each input can be selectively AC or DC coupled. DC coupling allows both DC and AC signals through, while AC coupling accepts only AC signals. Figure 4.22a shows a waveform containing both AC and DC. If the oscilloscope is DC coupled, then the waveform is displayed as drawn in Figure 4.22a. If the oscilloscope is AC coupled, then the DC portion of the waveform is blocked and only the AC portion is displayed, as shown in Figure 4.22b.

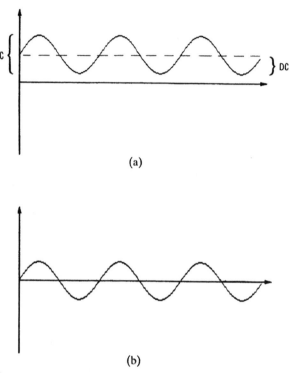

(a)

(b)

Figure 4.22   a) DC coupling causes the entire waveform to be displayed, including the DC portion. b) AC coupling removes the DC portion of the signal.

The previous example seems straightforward, but the issue of AC coupling may show up in other unexpected ways. Consider the pulse waveform in Figure 4.23a, shown as a DC-coupled scope would display it. This waveform appears to be a typical AC waveform so one might think that it would be unaffected by coupling. However, when the scope is AC coupled, the display does change. The waveform shifts down by about one-third of its original zero-to-peak value (Figure 4.23b). The original waveform did have some DC present in it (remember, DC is just the average value of the waveform). The AC coupling removed the DC, leaving a waveform whose average value is zero. Notice that the waveform is not centered exactly around zero volts, since its duty cycle is one-third. AC cou-

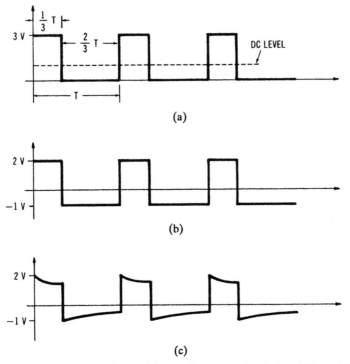

Figure 4.23 The effect of AC coupling on a pulse train. a) The original waveform displayed with DC coupling. b) The pulse train with DC component removed due to AC coupling. c) AC coupling may cause voltage droop due to the loss of low frequencies.

pling may also cause voltage "droop" or "sag" in the waveform (Figure 4.23c), due to the loss of low frequencies.

Many oscilloscopes have a convenient means of grounding the input (usually a switch near the connector). This is symptomatic of one of the most confusing things in using an analog scope—where is zero volts on the display? The ground switch allows the user to quickly ground the input and observe the flat trace on the display, which is now at zero volts. The line may then be set anywhere on the display that is convenient, using the display's position controls. Knowing where zero is defined along with the volts/division selection, determines the scale on the display. Fortunately, most digital scopes have eliminated the confusion by always displaying the data in a known calibrated manner indicated on the display.

4.12 Bandwidth Limit

Most scopes provide a *bandwidth limit* control, which activates a fixed frequency low-pass filter in the vertical amplifier. This has the effect of limiting the bandwidth of the scope (typically to about 20 MHz). Since bandwidth is such a desirable thing it may seem odd to intentionally limit it. Figure 4.24a shows an oscilloscope display of a sine wave with a noticeable amount of high frequency noise riding on it. When the bandwidth limit control is switched on (Figure 4.24b), the high frequency noise is eliminated, but the original sine wave remains uncorrupted. Of course, this works only when the interfering noise is mostly outside the bandwidth of the filter and the desired signal is inside the filter bandwidth.

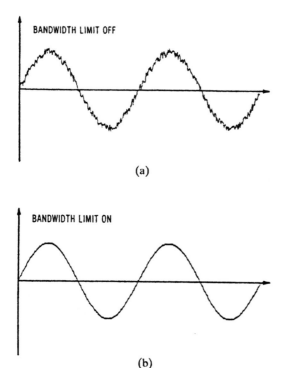

Figure 4.24   a) A noisy sine wave (with bandwidth limit control off). b) The same sine wave with the noise reduced by using bandwidth limit.

4.13   X-Y Display Mode

Most oscilloscopes have the ability to plot the voltage of one channel on the vertical (Y) axis and the voltage of the other channel on the horizontal (X) axis (Figure 4.25). This results in a voltage versus voltage display, usually called *channel versus channel* or *X-Y mode*. The timebase and triggering circuits are not used when operating in this manner.

This feature greatly enhances the usefulness of the oscilloscope since the horizontal axis is no longer limited to only time. Any other parameter that can be represented as a voltage can now be used as the X axis. More precisely, both the vertical and horizontal axes can be used to display any two parameters represented by voltages. For instance, if a current sensing resistor were used to convert a current into a voltage, a current could be plotted on

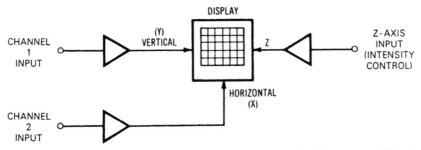

Figure 4.25 A simplified block diagram showing the X-Y display capability of an oscillo-scope. The Z-axis input can be used to control the intensity of the plotted waveform.

the vertical axis while another voltage is plotted along the horizontal axis. This and other X-Y applications are discussed in Chapter 5.

4.13.1 Z-Axis Input

The *Z-axis input* (also known as the *intensity modulation input*) provides a means for controlling the intensity of the display while in X-Y operation. The name "Z axis" comes from the fact that in addition to the X-axis and Y-axis information, the intensity of the display can be varied to provide an additional "axis" of information—the Z axis. If a positive voltage is applied to the Z-axis input, the trace is blanked (no intensity). If a negative voltage is applied to the Z-axis input, then the trace has full intensity. Voltages in between the two extremes produce less than full intensity (but not blanked). The actual voltages and even the polarity of the Z-axis input varies depending on the model of instrument. Given the proper Z-axis control signals, different sections of the trace can have different intensities. This is useful for highlighting a particular point on a display or for turning the trace off to start the plot over at a particular point (without having a trace drawn to that point).

4.14   High Impedance Inputs

The typical oscilloscope channel has a high-impedance input so that the circuit under test is not loaded significantly. The input can be modeled by a 1 MΩ resistor in parallel with a capacitance (Figure 4.26).[3] The value of the capacitance depends on the particular model of oscilloscope, but is generally in the range of 7 to 30 pF. The magnitude of a typical input impedance is plotted in Figure 4.27. At low frequencies, the capacitance acts like an open circuit and the impedance consists only of the 1 MΩ resistor. At about 8 kHz, the capaci-tor's impedance becomes significant as it just equals the 1 MΩ resistor impedance. The

---

3.   This is a simplified model that does not represent all of the impedance characteristics of an actual
       input, especially at higher frequencies.

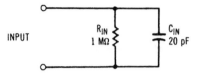

Figure 4.26   An RC circuit model for the input impedance of an oscilloscope.

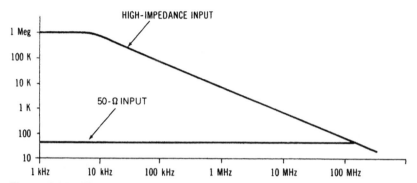

Figure 4.27   The magnitude of the input impedance for the high impedance and 50-Ω inputs.

impedance of the parallel combination continues to gradually decrease for higher frequencies. Although the input impedance is very high at low frequencies, it will tend to load the circuit being measured as the frequency increases. Remember, a 1 MΩ input is not 1 MΩ at high frequencies.

4.15 50-Ω Inputs

Higher bandwidth oscilloscopes offer a second type of input having a 50-Ω input impedance (Figure 4.28). It is usually the same connector as the high-impedance input, with a switch selecting between the two. This may be implemented by placing a 50-Ω resistor in parallel with the 1-MΩ input. Since 1 MΩ is much larger than 50 Ω, the effective parallel impedance is approximately 50 Ω. If a scope does not have the 50-Ω input built in, an appropriate external load can be placed in parallel with the high impedance input to produce the same result. The input impedance is modeled as a single 50-Ω resistor, with the input capacitance in parallel. In a 50-Ω system, the capacitive effect is less critical, resulting in a wider measurement bandwidth. Figure 4.27 shows that even though the 50-Ω input impedance starts out much smaller than the high-impedance input, it remains constant out to a higher frequency. The major disadvantage of the 50-Ω input is that it is too low of a load resistance for many circuits. (For these cases, very low capacitance active probes, which are designed to drive the 50-Ω input, are used to provide minimal circuit loading and greater overall bandwidth.) Of course, the 50-Ω input is especially convenient for systems that have an inherent 50-Ω impedance.

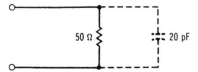

Figure 4.28   The circuit model for the 50-Ω scope input.

4.16   Digital Acquisition and Display Techniques

The architecture of a digital scope opens up some possibilities for processing the acquired waveform. With the waveform in digital form, the oscilloscope can apply signal processing and display algorithms to the waveform via a microprocessor or dedicated digital circuitry. This shift in technology has enabled new, powerful features to emerge in oscilloscopes.

4.16.1   Persistence

The digital scope can perform a *persistence* algorithm, which produces a display similar to the traditional analog storage scope. A typical digital persistence system works as follows. Each location on the display has a display intensity value stored in memory. When the screen location is first "hit" by a sample of the waveform, the intensity value is set to maximum. At some set interval in time, the intensity value is decremented. If this pixel location is not hit again by the waveform, the intensity level gradually reduces until it disappears from the display. If the pixel location is hit again, the process starts over with the intensity set to maximum. With a changing waveform, this causes the older waveform data to fade out while the new waveform data remains bright and visible. Rapidly varying waveforms cause some portions of the waveform to remain at high intensity (those that have been hit often) while other locations are dimmer because they are hit less often. This type of display gives the user information on how often certain voltage levels are hit by the waveform.

Of particular interest is the case where the persistence is infinite, which means the acquired samples are displayed indefinitely. *Infinite persistence* provides for worst-case characterization of signals by showing all variations in the waveform. Since persistence is implemented digitally, there is no problem with the display blooming as with analog storage scopes. One use of infinite persistence is the measurement of worst-case timing jitter on a signal (Figure 4.29). Other applications of infinite persistence are eye diagrams (as used in telecommunication channels) and detection of metastable states in logic design.

One problem with the use of infinite persistence is that the signal can get lost among the older waveform information. A slight variation on the display technique is to display the most recent waveform at full intensity, while showing older waveform samples at half intensity. This allows the oscilloscope user to view changes in the waveform while simultaneously keeping track of worst-case excursions.

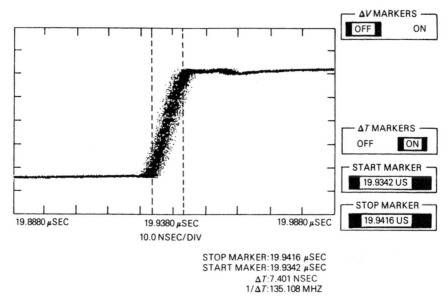

STOP MARKER: 19.9416 μSEC
START MAKER: 19.9342 μSEC
ΔT: 7.401 NSEC
1/ΔT: 135.108 MHZ

Figure 4.29 The time jitter present in a signal is measured using infinite persistence. The scope is configured to trigger on one edge of the waveform while viewing a later edge.

As color displays have become common in digital oscilloscopes, the persistence algorithms now employ color to show variations in the waveform, usually called *color-graded persistence*. Each pixel is displayed in a color that indicates how often the waveform happened to hit that pixel. Pixel locations that have been hit often are shown in one color, while less-hit pixels are shown in different colors. Again, this provides additional information to the scope user in terms of how the waveform varies over time.

4.16.2 Averaging

Averaging in a digital oscilloscope is performed by averaging together sample points associated with the same point on the waveform, but from different acquisitions. Averaging done in this fashion reduces the noise present in the measurement without bandwidth or rise time degradation (Figure 4.30).[4] In Figure 4.30, we can see that averaging has reduced the noise on the waveform while keeping the rise time of the signal unchanged. Compare this to the use of bandwidth limit, which can also be used to reduce the noise in a waveform but while sacrificing scope bandwidth.

---

4.   For averaging to remove the noise, the noise must be asynchronous to the trigger signal. Noise that is synchronous to the trigger will remain in the measurement.

Averaging has the side effect of slowing down the display's responsiveness. Usually the amount of averaging is selectable, with increased levels of averaging causing a more sluggish display. A benefit of averaging is increased vertical resolution as multiple waveform samples are averaged together. For example, a scope with an 8-bit ADC may have a vertical resolution associated with 12 bits, with a sufficient amount of averaging.

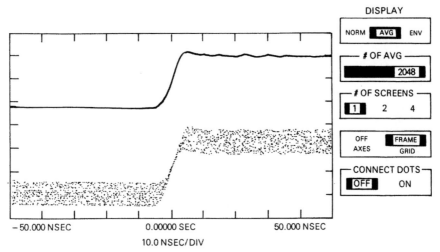

Figure 4.30   The noise present in the bottom trace is removed by averaging applied to the top trace.

4.16.3 Peak Detect

Many digital scopes on the market do not operate at their full sample rate on all sweep speeds, due to the limited size of the memory that holds the ADC samples. Let's consider the case where a scope has a sample rate of 1 GSa/sec (1 nsec per sample) and a memory depth of 100 k samples. When operating at full sample rate, this scope can capture 100 k × 1 nsec = 100 μsec of time. With 10 divisions across screen, this corresponds to 10 μsec/div. For timebase settings slower than 10 μsec/div, the sample rate of the scope must be reduced to keep from overrunning the memory. Most scopes handle this by decimating the samples, which is another way of saying "throwing some of the samples away." At 1 msec/div (or 10 msec total acquisition time), this scope would need to sample no faster than 10 MSa/sec. This reduction in sample rate means the scope may miss events in the waveform that would have normally been captured by the 1 GSa/sec ADC.

A larger memory for capturing the ADC samples would clearly help the situation, since any increase in memory size maps to increased sample rate at long acquisition times. Many scopes have a feature called *peak detect,* which addresses this problem using a different technique. With peak detect enabled, the scope keeps track of the minimum and maximum values in the sampled waveform during the time interval when the samples would normally have to be discarded. By storing the maximum and minimum samples, the scope can reliably capture events at the full sample rate of the scope, without overrunning the memory independent of timebase setting. Peak detection has the side effect of exaggerating any noise found on the waveform.

4.16.4 Roll Mode

The conventional oscilloscope display plots the waveform starting from the left and moving to the right. When the time/division is set to capture a short time window, the waveform quickly appears without giving the impression of being drawn left to right. However, on slower sweeps (e.g., 1 sec/division), the plotting of the waveform is quite noticeable.

Roll mode continually plots the waveform from the right side of the display, "rolling" the waveform to the left as new waveform samples become available. With roll mode, the scope user can monitor a slow-moving waveform, look for a particular waveform event, and then press stop to capture a complete view of the waveform. Because the new waveform data is always coming in on the right and shifting to the left, a full display of the most recent data is always present. With conventional "left-to-right" plotting, the display is often a mix of the previous acquisition and incoming waveform data.

4.16.5 Dual Timebase

Many scopes have the ability to present two different time scales simultaneously, which is used to present the "big picture" while zooming in on details of a waveform. This second timebase is usually called the *delayed timebase* or *delayed sweep*. This dual timebase operation is also sometimes known as *zoom* operation. Figure 4.31 shows how the delayed timebase can be used to zoom in on details of a pulse train.

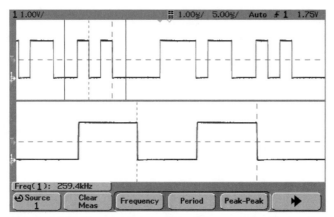

Figure 4.31 The upper waveform is the main timebase display showing a pulse stream. The lower waveform is the delayed timebase zoomed in on two pulses of the waveform. (Figure: Agilent Technologies. Reprinted with permission.)

The term "delayed timebase" comes from the original implementation in an analog scope, where a second timebase system (a complete set of analog timebase circuitry) was enabled after the main timebase started. Thus, the second timebase was inherently delayed from the main timebase. Usually the delayed timebase also had its own trigger system so that it could trigger on a particular edge of the waveform (as opposed to just waiting a preset amount of time). In digital scopes, the second "timebase" may or may not be a truly separate timebase. Often, it is the same acquired waveform displayed using a different time scale and is driven by the same trigger system as the main timebase.

4.16.6 Math Functions

In a digital scope, the waveform data is captured in digital form, which allows for math operations to be applied to the waveform. The most common math operations are addition, subtraction, multiplication, integration, and differentiation.

> ADD—The waveforms from two scope channels are added together (e.g., channel 1 + channel 2).
>
> SUBTRACT—The waveforms from two scope channels are subtracted (e.g., channel 1 − channel 2). This math function can be used to make differential measurements (Figure 4.32).
>
> MULTIPLY—The waveforms from two scope channels are multiplied together. This function is useful for displaying power versus time, when one channel represents voltage and the other channel represents current (via a current probe or other method).
>
> INTEGRATE—The integral of a scope waveform is displayed. Mathematically, this is defined as $f(t) = \int v(t)dt$.
>
> DIFFERENTIATE—The derivative of a scope waveform is displayed. Mathematically, this is defined as $f(t) = dv/dt$.
>
> FFT—The fast fourier transform (FFT) of the scope waveform is displayed. The FFT transforms the time domain waveform of the scope into a frequency domain display. The FFT function is described in more detail in Chapter 5.

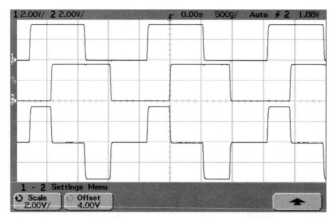

Figure 4.32 The two upper waveforms are channels 1 and 2. The lower waveform is a math function that subtracts channel 2 from channel 1 (i.e., channel 1 – channel 2), creating a differential measurement. (Figure: Agilent Technologies. Reprinted with permission.)

4.16.7 Cursors

Cursors are horizontal and vertical markers that allow the user to read out waveform data (both voltage and time) at any point on the waveform. Cursors are a powerful and versatile feature for measuring characteristics of a waveform. Usually, scopes have a pair of vertical cursors that read out two voltage points on the waveform and the difference between those two points, enabling convenient ΔV measurements. Similarly, a pair of horizontal cursors are usually provided that read out two points in time, along with the time difference between the two points, for Δt measurements. The reciprocal of the time difference (1/Δt) may also be displayed, as an aid in frequency measurement (Figure 4.33). To use this

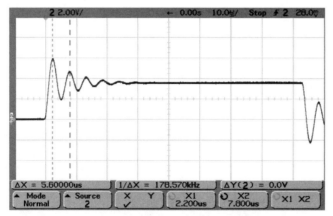

Figure 4.33 The ringing frequency of the pulse can be measured using the cursors. (Figure: Agilent Technologies. Reprinted with permission.)

feature, the user places the time cursors at each end of one period of the waveform, such that the frequency (1/t) is displayed.

4.17 Oscilloscope Specifications

A set of typical scope specifications is shown in Table 4.2. Bandwidth and rise time have already been covered in the general discussion in Chapter 1. The *deflection factor* indicates the vertical sensitivity settings that are available. In this example, settings between 1 mV and 5 volts per division are available in a 1-2-5 sequence (1 mV, 2 mV, 5 mV, 10 mV, 20 mV, 50 mV, etc.). Most scopes have a maximum deflection factor of 5 V per division, resulting in a vertical scale of 8 divisions or 40 volts. For higher voltage measurements, a 10:1 probe is used to extend the display range by a factor of 10, or 400 volts. The most sensitive range will depend on the particular scope, with values ranging from 1 to 5 mV per division.

The DC gain accuracy indicates the fundamental accuracy of the instrument. Oscilloscopes are normally specified for accuracy only at DC. It is assumed that the frequency response of the scope is flat at low frequencies and rolls off at higher frequencies. To achieve good pulse response (i.e., very little overshoot and ringing), the scope's response *must* roll off gradually with frequency. At its specified bandwidth, the scope's response is approximately 3 dB down from the DC response. Thus, a 1-volt peak-to-peak sine wave with a frequency near the scope's bandwidth will be measured as 0.707 volts.

The timebase settings indicate the time per division settings that are available on the scope. The timebase settings offered should correlate to the bandwidth and rise time of the scope. For example, Table 4.2 shows a bandwidth of 100 MHz and a rise time of 3.5 nsec with a minimum (fastest) timebase setting of 2 nsec/division. The fastest rising edge that the scope can display is roughly 3.5 nsec. With the timebase set to 2 nsec/division, the rising edge will be displayed as 1.75 divisions wide. If the scope had a faster timebase but not higher bandwidth, the rising edge of the fastest waveform would be spread out across more divisions. This provides no additional measurement accuracy or information, so scope manufacturers usually limit the fastest time per division setting to roughly twice the rise time of the scope.

The sample rate shown in Table 4.2 (200 MSa/sec) is twice the bandwidth of the scope (100 MHz). Most modern digital scopes have sample rates that exceed the bandwidth of the scope. Scopes that use random-repetitive or equivalent time sampling may have sample rates less than the bandwidth of the scope, but as technology improves these are becoming much less common, especially in the general-purpose scope market. Certainly, all other things being equal, more sample rate is better and some scopes offer sample rates that are 5 to 10 times their bandwidth. Note that the sample rate on the data sheet is usually a maximum sample rate, which may not be available on all timebase settings, especially for scopes that have limited memory depth.

Table 4.2   Abbreviated Specifications of a 100-MHz oscilloscope.

Specification	Value
Bandwidth (3 dB)	100 MHz
Rise time	3.5 nsec
Sample rate	200 MSa/sec
Memory depth	1 Meg
Vertical resolution	8 bits (0.39%)
Input impedance	1 MΩ, 14 pF
DC gain accuracy	± 2%
Deflection factor	1 mV/div to 5 V/div in 1-2-5 sequence
Timebase	2 nsec/div to 5 sec/div in 1-2-5 sequence

4.18   Scopesmanship

An oscilloscope is a relatively complex instrument and can be intimidating to the first-time or infrequent user. The large number of controls on the user interface can be confusing and frustrating for a novice user. A few comments are in order to help the first-time user get started.

After carefully reading the manufacturer's operating manual, the best way to get started with a measurement is to put the oscilloscope into a known state, which will at least get something on the display. Most digital scopes offer a feature called *autoscale* or *autosetup,* which automatically evaluates the waveform, chooses an appropriate trigger condition, and selects reasonable horizontal and vertical scales. Autoscale may not produce the perfect setup for a particular waveform or application, but it will generally get the signal onscreen so the user can manually zero in on the right measurement. Pushing this one button can prevent a lot of user frustration.

Assuming that autoscale is not available, a suggested starting point for an oscilloscope measurement is listed in Table 4.3. This can only be a start, as each measurement is somewhat different.

Hopefully, a trace will appear on the display after the oscilloscope is set up and the scope settings can be optimized for the particular measurement. If there is no trace at all, then things like the power switch (yes, the power switch!) and intensity control should be checked. Perhaps the waveform is just offscreen because it is much larger than the volts/division setting will allow onscreen. Try grounding the input—a horizontal line corresponding to zero volts should appear. Some analog scopes have a *beam finder* button that, when pushed, gives the user some idea where an offscreen trace is hiding.

Table 4.3 Suggested Starting Point for Oscilloscope Measurements

Control	Starting Setting
Vertical amplifier	Input coupling: DC volts/div: 1 volt or 1/4 of the expected zero-to-peak voltage
Timebase	Timebase operation (not XY or other mode) Sweep: Auto Time/div: 1 msec or 1/(expected frequency)
Trigger	Internal trigger (viewed channel) Trigger level: 0 volts Trigger slope: positive Trigger coupling: DC

If the trace is on screen, but is not stable, then the triggering controls should be adjusted. Try tweaking the trigger level to make the waveform stable. The slope and trigger coupling may also be helpful. If the display is stable but is scaled improperly, adjust the time/division and/or the volts/division knobs.

Probably the best advice for operating an oscilloscope is *try something carefully*. The two approaches that do not work are (1) just sitting in front of the instrument, staring at it and (2) twisting every knob until all of the controls are guaranteed to be in the wrong position. Instead, the user should make an educated guess as to what control might fix the problem and try it. If it does not improve the situation, the control should probably be returned to the original setting. Try not to get the oscilloscope so fouled up that only a seasoned technician can straighten it out. If in doubt, revert back to the suggested starting point.[5]

4.19 Mixed Signal Oscilloscope

With most scopes having either two or four input channels, it is easy to run out of channels when making measurements, especially when there is significant digital circuitry in the system. For purely digital circuits, a logic analyzer, which usually has 32 or more channels, can be employed. However, to measure analog signals or the analog characteristics of digital signals, a scope is required.

Agilent Technologies has addressed this need for measuring mixed analog/digital systems with the *mixed signal oscilloscope*. The mixed signal oscilloscope (MSO) can be thought of as a general-purpose scope that has additional channels added to it that are designed to be used only with digital signals. Figure 4.34 shows an MSO that has 2 analog

---

5.   Another approach that works is to go find "Joe." Joe is a mythical person that exists in all major electronics companies who has extensive knowledge of instruments and can always figure out how to set up a measurement.

(scope) channels and 16 digital (timing) channels. These timing channels are basically the same as the input channels found on logic analyzers (see Chapter 9). That is, they are designed to simply indicate whether a digital signal is high or low at any point in time. They are essentially scope channels with "one bit" of vertical resolution and limited voltage range.

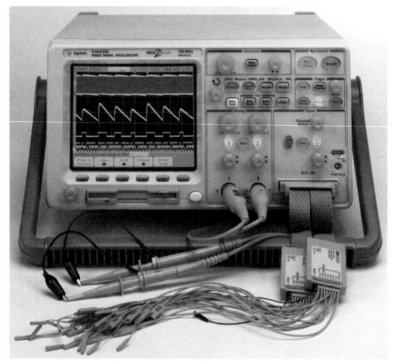

Figure 4.34 A mixed signal oscilloscope adds logic timing channels to an oscilloscope to provide better viewing and triggering when working with mixed analog/digital circuits.

This instrument operates primarily as a scope but with the added capability of 16 logic timing channels. Together, up to 18 channels of information can be viewed at one time, providing a more complete look at the operation of a mixed-signal circuit than with a conventional scope.

A typical mixed-signal measurement is shown in Figure 4.35a. The top two waveforms are two analog signals, captured by the two scope channels. The first signal is a digital signal with varying pulse width, while the second signal is the low-pass filtered version of the first. The lower half of the display shows the timing waveforms from eight of the digital logic channels, which are used to control the pulse width. The same measurement is shown in Figure 4.35b but with a faster timebase setting, which expands the waveform and reveals additional detail of the waveform.

Table 4.4   Abbreviated Specifications of a Mixed-Signal Oscilloscope

Specification	Value
Number of channels	Analog: 2 Digital: 16
Timebase	2 nsec/div to 5 sec/div in 1-2-5 sequence
Analog Channels	
Bandwidth (3 dB)	100 MHz
Rise time	3.5 nsec
Sample rate	200 MSa/sec
Memory depth	1 Meg
Vertical resolution	8 bits (0.39%)
Memory depth	1 Meg
Input impedance	1 MΩ, 14 pF
DC gain accuracy	± 2%
Deflection factor	1 mV/div to 5 V/div in 1-2-5 sequence
Digital Channels	
Number of channels	16
Timing resolution	2.5 nsec (400 MHz)
Memory depth	2 Meg
Trigger sequence levels	2
Maximum input voltage	± 40 V peak
Minimum voltage swing	500 mV$_{P-P}$
Voltage threshold range	−6 V to +6 V in 10 mV increments

a)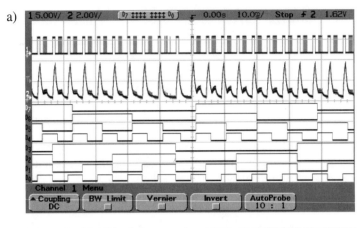

b)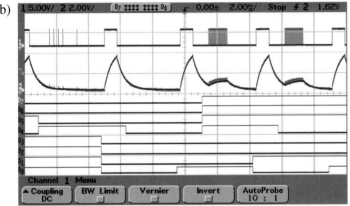

Figure 4.35   a) A measurement example of mixed analog and digital signals. b) The same measurement on a faster timebase setting shows more waveform detail.

4.20   Oscilloscope Probes

The high impedance input of the oscilloscope sometimes can be connected directly to the circuit under test using a simple cable. It is highly recommended that any cable used be shielded in order to minimize noise pickup. However, for maximum performance an oscilloscope probe matched to the input of the scope should be used.

4.20.1   1X Probes

1X probes, also known as *1:1* (one-to-one) *probes*, simply connect the high impedance input of the oscilloscope to the circuit being measured. They are designed for minimum loss and easy connection but otherwise they are equivalent to using a cable to connect the scope. Figure 4.36 shows the circuit diagram for a high impedance scope input connected to a circuit under test. The circuit under test is modeled as a voltage source with a

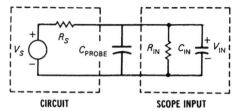

CIRCUIT   **SCOPE INPUT**

Figure 4.36 Equivalent circuit for when the high-impedance input is connected to a circuit using a 1X probe.

series resistor. The 1X probe will introduce a significant amount of capacitance, which appears in parallel with the input of the scope. A 1X probe may have around 40 to 60 pF of capacitance, which is usually larger than the oscilloscope input capacitance.

4.20.2 Loading Effects

The impedance of the circuit and the input impedance of the oscilloscope together produce a low-pass filter. For very low frequencies, the capacitor acts as an open circuit and has little or no effect on the measurement. For high frequencies, the capacitor's impedance becomes significant and loads down the voltage seen by the oscilloscope. Figure 4.37 shows this effect in the frequency domain. If the input is a sine wave, the amplitude tends to decrease and the phase is shifted with increasing frequency.

The loading also affects the oscilloscope's response to a step change in voltage. The loading due to the input impedance of the scope (and the probe capacitance) can be broken into two parts: *resistive loading* and *capacitive loading*. Figure 4.38 shows the probe and scope input loading broken into resistive and capacitive loading, which can be analyzed independently. The resistive loading is due entirely to the input resistance of the scope, while the capacitive loading is due to the probe capacitance combined with the scope input capacitance.

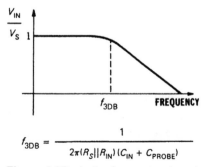

$$f_{3DB} = \frac{1}{2\pi (R_S || R_{IN})(C_{IN} + C_{PROBE})}$$

Figure 4.37 In the frequency domain, the response of the 1X probe rolls off at the higher frequencies due to capacitive loading.

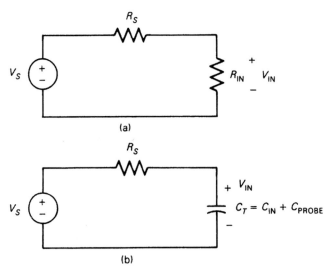

Figure 4.38 The loading of a circuit can be divided into (a) resistive loading and (b) capacitive loading.

The resistive loading circuit of Figure 4.38a is another example of the voltage divider circuit. Thus, the voltage delivered to the scope input, V_{IN}, is a replica of V_S but with reduced amplitude. For a voltage step from zero to V_{MAX} at time t = 0,

$$v_{IN}(t) = V_{MAX}\left[\frac{R_{IN}}{R_{IN} + R_S}\right] \qquad for \ t > 0$$

The effect of capacitive loading is more complex and results in an exponential response in the voltage. For a V_S voltage step, which goes from zero volts to V_{MAX} volts, V_{IN} obeys the following equation (see Appendix B for the complete analysis).

$$v_{IN}(t) = V_{MAX}\left[1 - e^{-t/(R_S C_T)}\right] \qquad for \ t > 0$$

The step response due to the two loading effects are shown in Figure 4.39. The resistive loading changes the size of the voltage step, but does not change the waveform shape. Capacitive loading slows down the rise time of the step but eventually settles out to the same final value as the ideal response. As shown in Chapter 1, the bandwidth and rise time of a system are inversely related. Since the bandwidth of the instrument is effectively being decreased, the rise and fall times of pulse inputs will be increased.

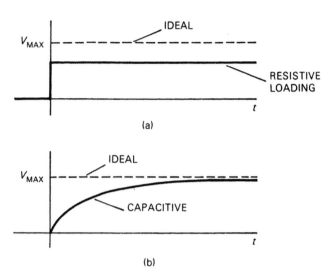

Figure 4.39 Resistive loading (a) changes the voltage level of a step while capacitive loading (b) causes the rise time to increase.

The circuit model used for this analysis may not be accurate for all types of practical circuits. The output resistance (drive capability) of digital circuits may vary with the output voltage and cause the loading effect to be different. Even though this model is not 100% accurate for such a circuit, the basic principles of resistive and capacitive loading still apply. This means that load capacitance will slow down the rise time of the signal while resistive loading will tend to change the output amplitude. Increased rise time in a digital circuit is translated into increased delay when the signal reaches the next logic gate. This is because it will take longer for the signal to rise to the logic threshold, causing the next gate to switch later. The 1 MΩ input impedance of the typical oscilloscope is large enough to prevent resistive loading of most digital circuits, but the capacitive loading of a 1:1 probe will introduce significant delay into the signal.

4.20.3 10X Probes

10X probes (also called *10:1 probes, divider probes,* or *attenuating probes*) have a resistor and capacitor (in parallel) inserted into the probe. Figure 4.40 shows the circuit for the 10X probe connected to a high impedance input of an oscilloscope. If $R_1 C_1 = R_2 C_2$ then this circuit has the amazing result that the effect of both capacitors exactly cancel. In practice, this condition may not be met exactly but can be approximated. The capacitor C_1 may be made adjustable or an additional adjustment capacitor, C_A, can be added to enable a near-perfect match. Under these conditions, the relationship of V_S to V_{IN} is:

$$V_{IN} = V_S R_2/(R_1 + R_2)$$

which should be reminiscent of the voltage divider equation. R_2 is the input resistance of the scope's high-input impedance (1 MΩ) and $R_1 = 9R_2$. From the previous equation, this results in

$$V_{IN} = (1/10)\ V_S$$

So the net result is a probe and scope input combination that has a much wider bandwidth than the 1X probe, due to the effective cancellation of the two capacitors. The penalty that is incurred is the loss of voltage. The oscilloscope now sees only one-tenth of the original signal voltage (hence the name 10X probe). Also notice that the circuit being measured sees a load impedance of $R_1 + R_2 = 10$ MΩ, which is much higher than with the 1X probe.

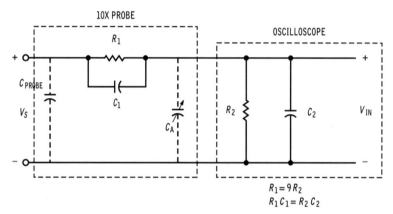

Figure 4.40 Circuit diagram showing a 10X probe used with the high impedance input. The effect of the capacitors cancel when C_A is adjusted properly.

With a 10X probe, both the resistive and capacitive loading effects are reduced (relative to a 1X probe).[6] Although the input capacitance of the scope is ideally cancelled, there is a remaining capacitance due to the probe, C_{PROBE}. This capacitance, which is specified by the manufacturer, will load the circuit under test.

The factor of 10 loss in voltage must be considered when choosing a probe. For measuring large signals, the loss is not a problem but measuring low-level signals will be more difficult. The effective sensitivity of the scope is reduced by a factor of 10. A scope that has a minimum vertical deflection factor of 2 mV per division will behave like it has 20 mV per division when using a 10X probe.

---

6.   Some 10X probes have a resistor across the probe input such that the resistive loading is 1 MΩ. These probes do not represent an improvement in resistive loading over the 1X probe, but they do have less capacitive loading.

Most modern oscilloscopes automatically sense the probe factor of the probe being used and adjust the onscreen readings and settings appropriately. The approach is to have an electrical contact on the probe connector that indicates the probe factor that should be used. (Typically, this information is encoded as a resistance value but other schemes can be used.) If the oscilloscope or the probe being used doesn't have automatic probe factor sensing, the user must either enter the probe factor into the scope or must account for the factor when interpreting the results.

Some probes are designed to be conveniently switched between 1X and 10X operation, providing the benefits of both probe configurations.

4.20.4 Other Attenuating Probes

Other types of attenuating or divider probes are available including 50:1 and 100:1 probes. The general principles of these probes are the same as the 10X divider probe: voltage level and bandwidth are traded off. To obtain wider bandwidth, more loss is incurred in the probe and less voltage is supplied to the input of the scope. This may require a more sensitive scope for low-level measurements. Some divider probes use the scope's 50-Ω input instead of the 1-MΩ input.

4.21 Probe Compensation

To maximize the bandwidth of the 10X probe, the probe capacitor must be adjusted precisely such that the input capacitance of the scope is cancelled. This is accomplished by a procedure known as *compensation.*

The scope probe is connected to a square wave source called the *calibrator,* which is built in to the scope. The probe is then adjusted to make the square wave as square and flat-topped as possible. Figures 4.41a and 4.41b show the oscilloscope display during compensation with an overcompensated and undercompensated probe. Figure 4.41c shows the display when the probe is properly compensated.

As discussed in Chapter 1, the square wave is a wide bandwidth signal rich in harmonics. If the probe is adjusted so that it measures a square wave with a minimum of waveform distortion, then the probe will be correctly compensated for wide bandwidth signals in general.

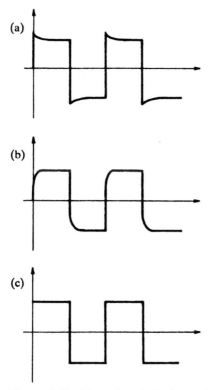

Figure 4.41 Examples of divider probe compensation: (a) overcompensated, (b) under-compensated, and (c) properly compensated.

4.22 Active Probes

So far, all of the probes discussed have been simple passive circuits with no active components such as transistors or integrated circuits. In cases where extremely low capacitance and wide bandwidth are required for high-frequency measurements, an *active probe* may be used. An active probe has a small wide-band amplifier built into it near the tip of the probe that is designed to have very little capacitance on its input. The output of the amplifier is usually matched to drive the 50-Ω input of the oscilloscope. This allows a length of 50-Ω cable to be used between probe and scope without any additional capacitive loading effects.

Active probes (Figure 4.42) require a power source for the active circuitry. This power may be supplied from the scope via the probe interface or from an external power supply box. Active probes that are powered directly from the scope are more convenient to use but the scope and probe must be designed to work together. The external power supply box generally allows an active probe to be used with any scope and other instruments.

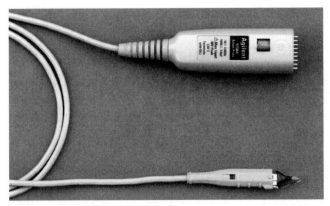

Figure 4.42   A compact active probe with a bandwidth of 4 GHz. (Photo: Agilent Technologies. Reprinted with permission.)

4.23   Differential Measurements

Most oscilloscope inputs are BNC connectors with one side directly connected to the instrument chassis (which is in turn connected to the power line safety ground through the power cord). This is the most practical and economical way to build the instrument. Many measurements are made relative to system common, which is often connected to ground, in which case having the scope input grounded is not a problem. For some measurements, however, it is desirable to connect the scope input to arbitrary points in the circuit, including ones that are not at ground. Attaching a standard scope probe's ground to such a point will force that point to ground. Excessive current may flow, ruining the circuit or the measurement.

Some scopes have floating or differential inputs that allow both leads of the input to be connected away from ground. In this case, the grounding problem is avoided. In particular, battery-powered, portable scopes *may* have the ability to float away from ground since they don't have a direct power line connection. *Be sure to check the instrument operating manual to understand under what conditions it is safe to have the scope floating away from ground since this is a critical safety issue.*

Another way to deal with the grounding problem is to use a differential measurement technique using two channels of the scope. A differential measurement is created by setting the scope to display Channel 1–Channel 2. Channel 1 is connected to the point in the circuit taken to be the more positive voltage. Channel 2 is connected to the other voltage point and the oscilloscope ground is connected to the circuit ground. Thus, the scope displays the difference between the two voltage points with neither one required to be at ground.

4.23.1   Differential Probe

A *differential probe* eliminates this problem by providing two scope probe inputs, which can be floating relative to the scope's chassis. The output voltage of the probe is the difference between the voltages on the two input terminals, allowing it to drive the chassis-referenced input of an oscilloscope. A differential probe has an amplifier built into it that produces a signal that is the difference between the two input voltages. Like the active probe, the differential probe must have power supplied to it to work. The probe power may be supplied by an internal battery, an external power supply, or the scope.

The differential amplification is not perfect and the error is specified in terms of *common mode rejection ratio* (CMRR). To measure CMRR, both inputs are driven with the same signal. Ideally, the output (which is the difference between the two inputs) is always zero. But in a real probe there is some small output voltage.

$$\text{CMRR} = \frac{\text{Input voltage (both inputs driven simultaneously)}}{\text{Output voltage}}$$

Typically, the CMRR of a differential probe is best at low frequencies and degrades at higher frequencies. Differential probes vary considerably in their performance. Some are designed for low-frequency measurements with a bandwidth of around 20 MHz. At the other extreme are probes intended for measuring differential high-speed digital signals, requiring a probe bandwidth on the order of 1 GHz.

4.24   High-Voltage Probe

Most conventional oscilloscope probes are rated to withstand operating voltages of around 450 volts (DC plus peak AC). For measuring higher voltages, a specialized *high-voltage probe* must be used. High-voltage probes are typically rated from 5 kV to 30 kV (maximum) and have an attenuation factor of 1,000. Such probes are physically large compared to standard probes to prevent high-voltage arcing. A typical high-voltage probe is shown in Figure 4.43.

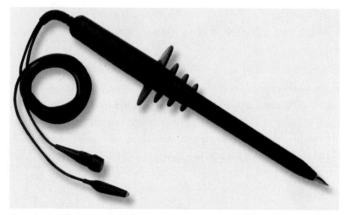

Figure 4.43   A typical high-voltage probe. (Photo: Agilent Technologies. Reprinted with permission.)

4.25   Current Probes

Oscilloscopes are designed for a voltage input, but can be used to measure current using a *current probe*. A current probe (Figure 4.44) has a set of jaws that enclose the wire that the measured current is flowing through. No electrical connection is needed. The circuit does not have to be broken or altered in any way, as the current probe measures whatever current is passing through its closed jaws.

Current probes generally use one of two technologies. The simplest uses the principle of a transformer, with one winding of the transformer being the measured wire. Since transformers work with only AC voltages and currents, current probes of this type do not measure direct current.

The other type of current probe works using the principle of the *hall effect*. The hall effect produces an electric field in response to a current present in an applied magnetic field.[7] This technique requires the use of an external power supply, but does measure both alternating and direct current (AC and DC).

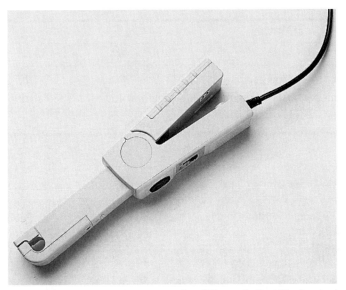

Figure 4.44   A typical current probe. (Photo: Agilent Technologies. Reprinted with permission.)

Since current probes measure the current enclosed by their jaws, several techniques can be used that are unique to the current probe. If the sensitivity of the probe and oscilloscope combination is too low for a particular measurement to be made, several turns of the

---

7.   For more information on the hall effect, see Halliday, Resnick, and Crane, 2001.

current-carrying wire can be inserted into the jaws. The probe will effectively have a larger current to measure (the original current times the number of turns). In a similar manner, the difference between two currents can be measured if the two wires in question are inserted, but with the currents flowing in opposite directions (the sum will be measured if the currents are flowing in the same direction). Of course, the physical size of the wires and the current probe will be a factor in determining how many wires can be inserted. Although the current does not require a direct electrical connection, it still removes energy from the circuit under test. Normally, this small amount of energy loss will not disturb the circuit, but can be a factor in some cases.

4.25.1 Probe Specifications

Table 4.5 summarizes the typical specifications of the most common scope probes that have been discussed. Actual characteristics will vary according to manufacturer and model.

Table 4.5   Typical Specifications of Oscilloscope Probes

Probe Type	Bandwidth	Resistive Load	Capacitive Load
1X passive	20 MHz	1 MΩ	70 pF
10X passive	100 MHz	10 MΩ	15 pF
10X passive	500 MHz	10 MΩ	9 pF
Active probe	2.5 GHz	100 kΩ	0.8 pF
High-voltage probe	1 MHz	500 MΩ	3 pF

4.26 References

"Agilent 54600-Series Oscilloscopes User's Guide," Agilent Technologies, Publication No. 54622-97002, March 2000.

"Agilent Technologies 54600-Series Oscilloscopes Data Sheet," Agilent Technologies, Publication No. 5968-8152, April 2000.

"Agilent Technologies 54810/15/20/25/35/45A Oscilloscopes Data Sheet," Agilent Technologies, Publication No. 5968-8496, February 2000.

Halliday, David, Robert Resnick, and Kenneth S. Krane. *Physics*, 5th ed., New York: Wiley, 2001.

Stanley, William D., Gary R. Dougherty, and Ray Dougherty. *Digital Signal Processing*, 2nd ed., Reston, VA: Reston, 1984.

"Understanding and Minimizing Probing Effects," Hewlett-Packard Company, Application Note 1210-2, Publication No. 5091-1800E, 1991.

Wedlock, Bruce D., and James K. Roberge. *Electronic Components and Measurements*. Englewood Cliffs, NJ: Prentice Hall, 1969.

Oscilloscope Measurements

T he oscilloscope produces a picture of the voltage waveform that is being measured. This allows the instrument user to think in terms of the actual waveform when making measurements. Therefore, determining zero-to-peak voltages, RMS voltages, and the like simply means interpreting the scope display using the theory in Chapter 1.

The oscilloscope is a very flexible and capable instrument that can be applied to a wide variety of electronic measurement needs. In this chapter, we explore how to apply the oscilloscope, alone and in concert with other instruments.

5.1 Basic Waveform Measurements

The primary role of an oscilloscope is to enable the user to view a voltage waveform. In this role, the amplitude, frequency, shape, and quality of the waveform is revealed. Specific parameters of the waveform such as amplitude, frequency, and period can be measured either by eyeballing the waveform relative to the graticule or by automated measurements in the scope.

5.1.1 Sine Wave Measurements

Figure 5.1 shows a typical display of a sine wave using an oscilloscope. The usual sine wave parameters can be determined from the display with some care. The peak-to-peak voltage can be found first in terms of display divisions and then converted to volts. The peak-to-peak value of the sine wave in Figure 5.1 is four divisions. If the vertical sensitivity

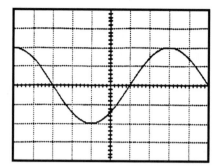

Figure 5.1 A sine wave voltage being measured with an oscilloscope. If the vertical sensitivity is 0.5 volts/division, then V_{P-P} = 2 volts and V_{0-P} = 1 volt.

is set to 0.5 volts per division, then the peak-to-peak voltage is $4 \times 0.5 = 2$ volts.[1] The zero-to-peak voltage is just two divisions, so $2 \times 0.5 = 1$ volt.

The RMS voltage is not as easy to determine, at least not directly from the waveform display. We can use the known relationship between the zero-to-peak and RMS values for a sine wave.

$$V_{RMS} = 0.707 \, V_{0-P} = (0.707)(1 \text{ volt}) = 0.707 \text{ volts RMS}$$

This calculation is valid only for a sine wave (conversion factors for other waveforms were covered in Chapter 1). To be precise, the oscilloscope cannot measure RMS voltage directly, but does give the user enough information to compute the value for simple waveforms.

The above measurements used the vertical axis to determine voltage information. The horizontal (or time) axis can be used to determine the period of the waveform. The period of the waveform in Figure 5.1 is eight divisions. With the horizontal axis set at 200 μsec/div, the period of the signal is $(8 \times 200 \text{ μsec}) = 1.6$ msec. Although the frequency cannot be read from the oscilloscope directly, it can be computed using the period of the waveform.

$$f = 1/T$$

The frequency of the waveform in the figure equals 1/1.6 msec or 625 Hz. Frequency is another parameter that cannot be determined directly viewing the waveform but the scope gives us sufficient information to compute the value.

---

1.   This value might need to be adjusted for the probe attenuation factor if an attenuating probe is used. See Chapter 4.

> **Example 5.1**
>
> **Determine the zero-to-peak voltage, the peak-to-peak voltage, period, and frequency of the waveform shown in Figure 5.2.**
>
> The waveform is 1.5 divisions zero-to-peak and 3 divisions peak-to-peak.
>
> V_{0-P} = 1.5 div × 0.2 volts/div = 0.3 volts
>
> V_{P-P} = 3 div × 0.2 volts/div = 0.6 volts
>
> The period of the waveform is 3 divisions.
>
> T = 3 div × 500 μsec/div = 1.5 msec
>
> f = 1/T = 666.67 Hz

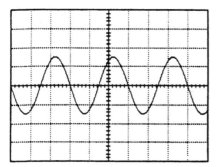

Figure 5.2 Oscilloscope waveform for Example 5.1. Vertical sensitivity = 0.2 volts/division and timebase = 500 μsec/division.

5.1.2   Oscilloscopes and Voltmeters

Figure 5.3 shows a voltmeter and an oscilloscope connected to the output of a function generator. A comparison of the two measurements will highlight the advantages of each. The oscilloscope provides a complete representation of the waveform out of the function generator. Its peak-to-peak and zero-to-peak AC values are easily read and for most waveforms the RMS value can be computed. If there is some DC present along with the waveform, then this, too, will be evident in the voltage versus time display (assuming that the scope is DC coupled). The period and frequency of the waveform can also be measured.

The voltmeter supplies only voltage information about the waveform. Most voltmeters read RMS directly, but the accuracy of the reading may be dependent on the waveform shape if the meter is an average-responding type. Also, no indication is given as to the actual shape of the waveform. The user may assume a given shape, but distortion due to improper circuit operation may cause the waveform to be substantially different.

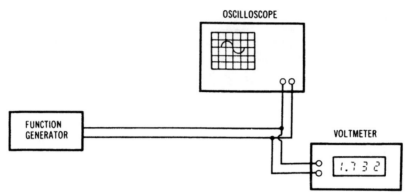

Figure 5.3   The output of a function generator is measured by an oscilloscope and a voltmeter.

The voltmeter is easier to use. It requires less interpretation of its display, has fewer controls to adjust, and its physical size is usually smaller and more convenient than the scope. Voltmeters are also generally much more accurate for amplitude measurements than oscilloscopes. A typical oscilloscope amplitude accuracy is around 2 percent, while voltmeter accuracy is usually much better than 1 percent. With a True-RMS voltmeter, an accurate RMS value can be read directly from the meter.

5.1.3   Square Wave and Pulse Measurements

A basic sine wave with low distortion makes for easy measurement interpretation. The maximum voltage, the minimum voltage, and the period completely describe the waveform. The situation gets a bit more complex when we consider square wave or pulse waveforms, especially with nonideal waveshapes. Figure 5.4 shows how the standard waveform parameters are defined on pulse waveforms that are not perfectly shaped. These measurements can be performed using the cursors of the oscilloscope, or these parameters are measured automatically with digital scopes.

The maximum and minimum values of the waveform are easy to identify, but note that the maximum is not necessarily the same as the "high level" of the waveform due to overshoot. Similarly, the minimum value is not always the same as the "low level." The high level and low levels of the waveform are labeled *top* and *base* in Figure 5.4. The *peak-to-peak* value is defined as the maximum minus the minimum, while the *amplitude* of the waveform is defined as *top* minus *base*. *Overshoot* and *preshoot* are measures of how far the waveform exceeds the *top* and *base*.

The timing parameters of the waveform are also affected by the imperfections in the waveform. Finite rise and fall times force us to define the voltage at which the timing measurements are taken. As shown in Figure 5.4, the *period* and *pulse width* (positive and nega-

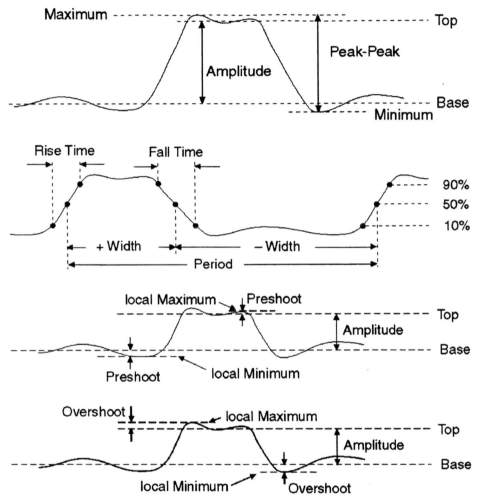

Figure 5.4 Definition of waveform measurements for pulse waveforms. (Figure: Agilent Technologies. Reprinted with permission.)

tive) are measured at the 50% points on the waveform (50% of the difference between *top* and *base*), while the *rise time* and *fall time* are taken at the 10% and 90% points.[2]

The *frequency* of the waveform is calculated as the reciprocal of the *period*.

$$f = 1/T$$

---

2. These are the most common definitions for pulse parameters. However, other definitions may be used such as the 20% and 80% points or absolute voltage levels that correspond to logic levels.

The *average* value of the waveform is calculated by averaging the voltage readings of the waveform over one period. (If a full period is not present, most scopes will compute the average over the entire displayed waveform.)

$$V_{average} = \frac{\sum_{i=1}^{N} v_i}{N}$$

The *RMS* value of the waveform is calculated by taking the square root of the average of the squares of each voltage reading in the waveform. (If a full period is not present, most scopes will compute the average over the entire displayed waveform.) Note that for waveforms that are pure DC, V_{RMS} equals the DC value of the waveform.

$$V_{RMS} = \frac{\sqrt{\sum_{i=1}^{N} v_i^2}}{N}$$

Automated measurements for frequency, average value, and RMS value are good examples of the analysis power in digital scopes, as these parameters require calculation and are difficult to read directly from the waveform.

5.2   Voltage Gain Measurement

One common measurement technique is to apply a sine wave to the input of a circuit and measure the resulting sine wave at the circuit's output (Figure 5.5). The circuit may increase or decrease the waveform amplitude and may also introduce a phase shift. The amplitudes and phases of the two sine waves are compared, giving the voltage gain and phase shift of the circuit.

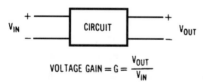

$$\text{VOLTAGE GAIN} = G = \frac{V_{OUT}}{V_{IN}}$$

Figure 5.5   The voltage gain of a circuit is determined by dividing the output voltage by the input voltage.

The input waveform is

$$v_{IN}(t) = V_{IN} \sin(2\pi f t)$$

and the output is

$$v_{OUT}(t) = V_{OUT}\sin(2\pi ft + \theta)$$

Note that the input waveform is defined such that it has a phase shift of zero.

It is often desirable to measure the *gain* of circuits such as amplifiers, filters, and attenuators. The voltage gain is defined as

$$Voltage\ gain = G = V_{OUT}/V_{IN}$$

where V_{IN} and V_{OUT} are the voltages at the input and output of the circuit, respectively. In other words, gain describes how large the output is compared to the input. If the output is larger then the input, the gain is greater than one. If the output equals the input, then the gain is exactly one and if the output is less than the input, the gain is less than one.

As defined, V_{IN} and V_{OUT} are zero-to-peak voltages. However, the concept of gain can be applied using any type of voltage: DC, AC zero-to-peak, AC RMS, and so on, as long as both the input and output voltages are expressed consistently.

Circuits with gain less than one can be described as having a *loss*.

$$Voltage\ loss = V_{IN}/V_{OUT} = 1/G$$

5.2.1 Gain in Decibels

Voltage gain can also be expressed in dB:

$$Voltage\ gain(dB) = G_{dB} = 20 \cdot \log(V_{OUT}/V_{IN})$$

If the output is greater than the input, the gain (in dB) is a positive number. If the output equals the input, the gain is 0 dB and if the output is less than the input the gain is negative (in dB). The original definition of gain and loss showed that they were the inverse of each other (i.e., Voltage loss = 1/G). When expressed in decibels, the gain has the opposite sign as the loss. For example, a gain of –10 dB (output is actually smaller than the input) corresponds to a 10 dB loss.

5.2.2 AC Voltage Gain

Figure 5.6 shows a simple method for measuring AC voltage gain. A signal source is used to supply the input voltage and a two-channel oscilloscope is used to measure the input and output voltage of the circuit. The resulting oscilloscope display is shown in Figure 5.7. The values of the two voltages are determined from the display and the gain is calculated. If the signal level into the circuit is not critical, then it is desirable to set the sine wave source such that the value of V_{IN} is convenient for the gain calculation (e.g., set V_{IN} to 1 volt).

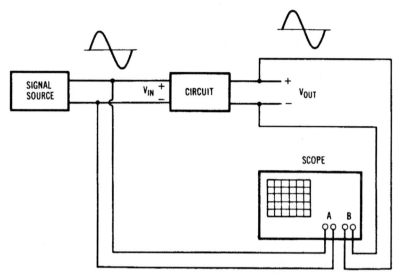

Figure 5.6   Instrument connections for measuring the voltage gain of a circuit.

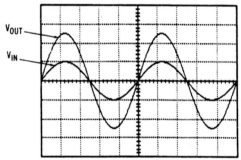

Figure 5.7   The oscilloscope display for measuring the gain of a circuit (vertical scale = 1 volt/division).

Again, the AC voltage can be described in a variety of ways but zero-to-peak and peak-to-peak are usually the most convenient when measuring with a scope.

Proper loading of the input and output should be maintained when making voltage gain measurements. Circuits that expect to be loaded with a particular impedance (Z_0 systems, for instance) should either be loaded with a resistor or an instrument with the appropriate input impedance. Also, the input to such a circuit should be driven with a source that has the correct output impedance.

Example 5.2

Determine the voltage gain for V_{IN} and V_{OUT} shown in Figure 5.7. Express the value in dB.

Figure 5.7 shows the zero-to-peak value of

V_{IN} = 1 div × 1 volt/div =1 volt zero-to-peak

and V_{OUT} = 2.5 div × 1 volt/div = 2.5 volts zero-to-peak

The voltage gain, G = V_{OUT}/V_{IN} = 2.5/1 = 2.5

Notice how easy the gain calculation was with V_{IN} equal to 1 volt.

In dB, G(dB) = 20 log 2.5 = 7.96 dB

5.3   Phase Measurement

The waveforms shown in Figure 5.7 have no phase difference between them, but this is not always the case. Many circuits introduce a phase shift between input and output. Depending on the particular application, the phase shift through a circuit can be an important parameter to be measured.

5.3.1   Timebase Method

The same setup shown in Figure 5.6 can be used to measure the phase shift through a circuit. If there is a nonzero phase shift through the circuit, the resulting oscilloscope display will look something like Figure 5.8. (The gain of the circuit is shown as 1 for simplicity.) The scope display gives the user a direct, side-by-side comparison of the two signals and the phase difference can be determined. First, the period of the sine wave is found in terms of graticule divisions. (Recall that one period corresponds to 360 degrees.) Next, the phase difference is determined in terms of graticule divisions. This can be done by choosing a convenient spot on one waveform and counting the divisions to the same spot on the other waveform. The starting edge of the sine wave (where it crosses zero) is usually a good reference point since most scope graticules have the middle of the display marked with the finest resolution. The phase difference in degrees can be computed using:

$$Phase\ shift(\deg) = \theta = 360 \cdot \frac{phase\ shift(divisions)}{period(divisions)}$$

Since interpreting the oscilloscope display is somewhat tedious, it is recommended that the user double-check the computed value.[3] This simply means to take another look at the dis-

---

3.   This is known as a reality check.

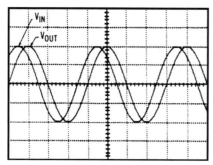

Figure 5.8 The phase difference between two sine waves can be measured using the timebase method.

play, roughly estimate the phase shift (knowing that one cycle is 360 degrees), and compare this to the calculated answer. The two should be roughly the same.

Phase can be positive or negative in value and determining the sign of the phase angle can be confusing. The waveform v_{OUT} is normally measured with respect to v_{IN}, so v_{IN} is the phase reference. If v_{OUT} is shifted to the left of v_{IN}, which means v_{OUT} leads v_{IN}, then v_{OUT} has a positive phase relative to v_{IN}. If v_{OUT} is shifted to the right of v_{IN}, then v_{OUT} lags v_{IN} and v_{OUT} has a negative phase angle. The use of the terms "lead" and "lag" can be helpful in describing phase relationships, as they may be less confusing than applying a sign to the phase angle.

Since phase repeats on every cycle (360 degrees), the same phase relationship can be described in numerous ways. For example, if v_{OUT} leads v_{IN} by 270 degrees, this is the same as v_{OUT} lagging v_{IN} by 90 degrees. Although both of these expressions are technically correct, common practice is to express phase differences within the range of ±180 degrees. In this example, the appropriate expression would be that v_{OUT} lags v_{IN} by 90 degrees.

Example 5.3

Determine the phase difference between V_{OUT} and V_{IN} as shown in Figure 5.8.

The period of both waveforms is 4 divisions. The phase shift in divisions is 1/2 of a division.

$$\theta = 360 \times 0.5/4 = 45 \text{ degrees}$$

Since V_{OUT} is shifted to the right of V_{IN}, V_{OUT} lags V_{IN} by 45 degrees or equivalently, V_{IN} leads V_{OUT} by 45 degrees.

5.3.2 Lissajous Method

Another method for measuring the phase between two signals is called the *Lissajous method* (or *Lissajous pattern*). Although somewhat more complicated, this method will usually result in a more accurate phase measurement. Figure 5.9 shows a scope connected such that the phase between the output and input of a circuit can be measured. The oscilloscope is configured in the X-Y mode with one signal connected to the horizontal input and the other signal connected to the vertical input.

Figure 5.10 shows the elliptical shape that results from this measurement. Two values, A and B, are determined from the scope display and are used to calculate the phase angle. The value A is the distance from the X axis to the point where the ellipse crosses the Y axis and the value B is the height of the ellipse, also measured from the X axis. The scope must be set up such that the ellipse is centered on the graticule, to provide for an accurate determination of A and B. The volts/division controls can be adjusted to allow convenient and accurate reading of the A and B values. Note that the two controls do *not* have to be set the same, since A and B are measured along the same axis.

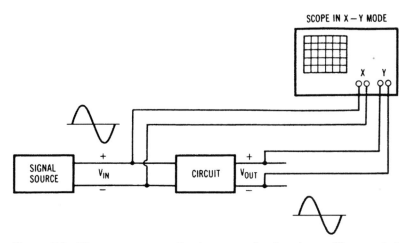

Figure 5.9 The proper connection for measuring the phase difference between the input and output of a circuit using the Lissajous method.

Two general cases must be considered. If the ellipse runs from lower left to upper right, then the phase angle is between 0 and 90 degrees (Figure 5.11). If the ellipse runs from lower right to upper left, then the angle is between 90 and 180 degrees. The angle can be computed from the A and B values using the appropriate equation, as shown in Figure 5.10. Unfortunately, the sign of the angle cannot be determined using this method. If the computed answer is 45 degrees, for example, the phase difference may be +45 degrees or –45 degrees. Said another way, the signal on the vertical axis may be leading or lagging the signal on the horizontal axis by 45 degrees. The time base method can be used to determine

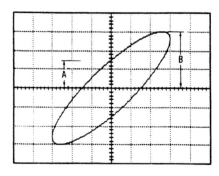
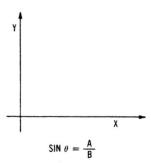

$$SIN\ \theta = \frac{A}{B}$$

Figure 5.10   The oscilloscope is operated in X-Y mode when using the Lissajous method.

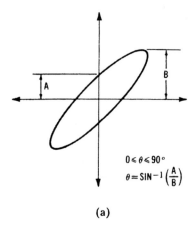

$$0 \leqslant \theta \leqslant 90°$$
$$\theta = SIN^{-1}\left(\frac{A}{B}\right)$$

(a)

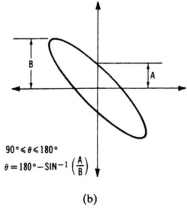

$$90° \leqslant \theta \leqslant 180°$$
$$\theta = 180° - SIN^{-1}\left(\frac{A}{B}\right)$$

(b)

Figure 5.11   The Lissajous method for computing phase has two general cases. (a) The ellipse runs from lower left to upper right. (b) The ellipse runs from upper left to lower right.

the sign, while using the Lissajous method for greater accuracy. A quick look using the time base method is also a good check on the results from the Lissajous method.

Three special cases of the Lissajous display are shown in Figure 5.12. When the ellipse collapses into a straight line, the two waveforms are in phase. This can be used as a very precise indication when adjusting for zero phase between two signals. If the ellipse is a perfect circle then the waveforms are exactly 90 degrees apart. Again, this could be plus or minus 90 degrees, with either signal leading or lagging. If the display becomes a straight line, but in the lower right/upper left orientation, then the two signals are exactly out of phase (180 degrees).

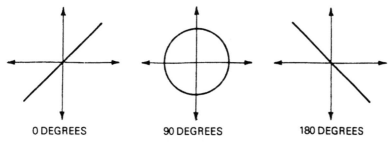

Figure 5.12   Three special cases of the Lissajous phase measurement.

Example 5.4

Determine the phase difference between the two signals given the Lissajous pattern shown in Figure 5.13.

The ellipse is lower left to upper right. First find the values for A and B.

A = 2.3 divisions

B = 3 divisions

$$\theta = \sin^{-1}(A / B) = \sin^{-1}(2.3/ 3) = 50 \text{ degrees}$$

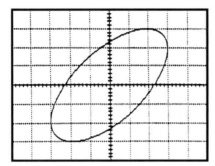

Figure 5.13   Lissajous pattern for Example 5.4.

5.4   Frequency Measurement (Lissajous Method)

The Lissajous method can also be used to compare the frequency of two sine waves. The oscilloscope, operating in X-Y mode, is connected as shown in Figure 5.14. The frequency being measured is connected to the vertical axis while the reference frequency (hopefully, precisely known) is connected to the horizontal axis. If the two frequencies are the same, then the situation is exactly the same as the phase measurement case. In Figure 5.15a, the oscilloscope display is shown for a frequency ratio of 1:1 and a phase shift of 90 degrees. If the phase is other than 90 degrees, then the display will not be a perfect circle, but an ellipse. Again, this case was covered under phase measurement. If the two frequencies are not quite exactly the same, then the display will not be stable and the ellipse will contort and rotate on the display.

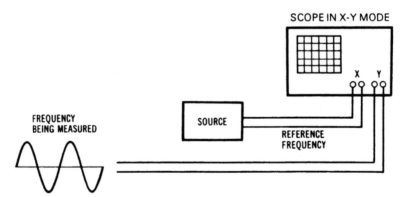

Figure 5.14   The Lissajous method can be used to compare an unknown frequency to a frequency reference using X-Y mode on the oscilloscope.

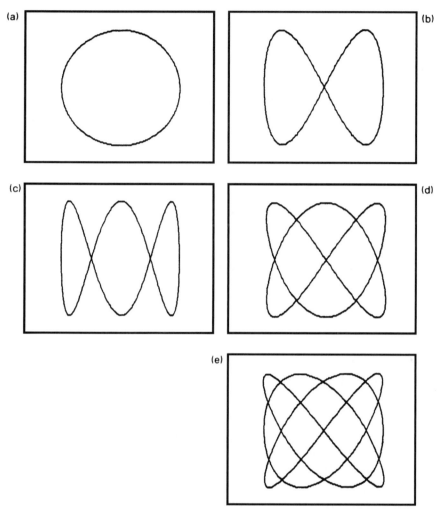

Figure 5.15 Typical Lissajous patterns for a variety of frequency ratios. The frequency ratio determined by the number of cusps across the top and sides of the pattern. (a) 1:1, (b) 2:1, (c) 3:1, (d) 3:2, (e) 4:3.

Other frequency ratios are also shown in Figure 5.15. They may also appear warped or slanted in different ways just like the ellipse is a slanted version of a circle for the 1:1 case. In general, the ratio of the two frequencies is determined by the number of cusps (or humps) on the top and side of the display. Consider Figure 5.15b. There are two cusps across the top and only one cusp along the side, therefore the frequency ratio is 2:1. In Figure 5.15d, there are three cusps across the top and two cusps along the side, resulting in a frequency ratio of 3:2. This technique can be applied to any integer frequency ratio. The display will be stable only when the frequency ratio is exact. In general, if the frequency sources are not phase locked together, there will be some residual phase drift between the two frequencies with a corresponding movement on the display.

This method of frequency measurement is clearly limited since it deals only with distinct frequency ratios. It does help if a sine wave source with variable frequency is available for use as the reference. Then the source can be adjusted so that the measured signal's frequency results in a convenient ratio. Since the method uses frequency ratios, the limiting factor in the accuracy of the measurement is the frequency accuracy and stability of the reference source. If the source used as a frequency reference is not more accurate than the oscilloscope timebase, then there is no advantage in using this method. Instead, the frequency should be computed from the period of the waveform measured in the timebase mode.

5.5 Digital Signal Measurement

As discussed in Chapter 1, digital signals can take on one of two valid states: HIGH or LOW. Assuming positive logic, these two states may be used to represent the binary numbers 1 and 0, respectively. Thus, most digital signals are pulsed in nature. In general, the pulse width and period will depend on the particular digital system. The digital signal may be periodic, repeating the same binary states in a predictable manner, or it may pulse in what appears to be a random manner, with no identifiable repetition. For instance, a digital signal on the data bus of a microprocessor will change according to the data and instructions being read from memory. Usually, there will be no discernible pattern to these binary numbers, and the digital signal will appear to be a sequence of random pulses. (In addition, the data bus may go into the high impedance state between transfers, confusing the situation even further.)

One should not expect to see perfectly clean pulses in digital signals. Even though digital systems are based on the concept of only two valid states, the actual voltage being observed is still an analog voltage capable of having any value within the power supply range. If the output of the driving circuit goes into the high impedance state, then the voltage may float to most any value and will be susceptible to cross talk from nearby signals. Also, the digital signal is subject to the same rise time limitations (due to the system bandwidth) as any other signal. Noise can also be present and, if large enough, can cause a digital signal to change state.

Oscilloscopes are an important tool for measuring digital signals, as they tend to show all of the imperfections of the signal. Sometimes this is unnecessary information that can be ignored, but in many cases it can be critical to understanding the circuit operation or malfunction. A logic analyzer is a better choice for viewing digital signals when the analog characteristics of the signal are known to be good. Logic analyzers have the advantage of handling large numbers of signals efficiently as well as providing very versatile trace capability. But for critical timing or signal quality measurements, the oscilloscope is the preferred test instrument.

5.5.1   Pulse Train

Digital logic signals can be thought of as a series of pulses or a pulse train, as shown in Figure 5.16. These pulses, pulse trains, and square waves can be measured and characterized using an oscilloscope. The measurement involves connecting the scope to the waveform of interest, obtaining a voltage versus time display of the waveform, and extracting the parameter of interest from the display. As discussed earlier in this chapter, digital scope automatic measurements may be used to quickly measure pulse parameters such as frequency, period, pulse width, and rise time. Of course, the oscilloscope bandwidth and rise time must be adequate so that significant measurement error is not introduced by the scope.

Various imperfections may exist in a pulse train, as shown in Figure 5.16. As covered earlier, real pulses have finite rise and fall times and overshoot or preshoot. The time it takes

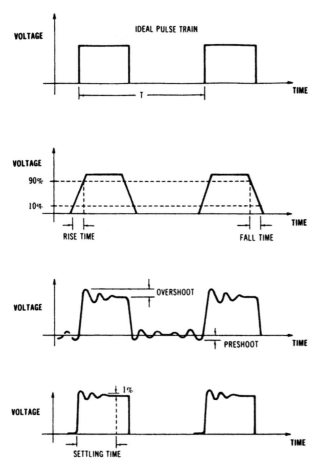

Figure 5.16   Digital signals can be thought of as ideal pulse trains, but they often have significant imperfections in practical circuit implementations.

for the pulse to settle out is called *settling time* and is measured as the time it takes to settle within some percentage of the final voltage, often 1%. The top of the pulse may not be perfectly flat, but have some small downward slope to it, called *droop* or *sag*. Abrupt voltage changes in the waveform are called *glitches*. Glitches are a common design problem in digital circuits, caused by cross talk between circuits or timing problems in the logic circuit. Glitches can be large enough to cause a digital signal to enter the undefined region or, in more extreme cases, to change logic state. Some circuits will tolerate a certain amount of glitches being present, but glitches are generally undesirable, as they may cause the circuit to fail, often intermittently.

5.5.2 Pulse Delay

The time delay between two digital pulses can be measured using a two-channel scope. Figure 5.17 shows two logic gates (inverters) connected end to end, driven by a pulse generator. In all logic gates, there is some delay between the input and output. The oscilloscope is set up to display the input and output of the two-gate circuit, such that the time

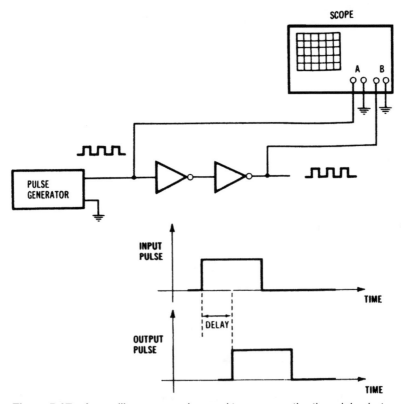

Figure 5.17 An oscilloscope can be used to measure the time delay between two pulses.

delay between the two waveforms can be measured. The time delay is found by determining the time between the two rising edges of the waveforms.

5.5.3 Serial Bit Stream

Digital information is often transferred one bit (binary digit) at a time in serial form over a single wire. This can significantly reduce the number of connections required to pass a binary number or series of numbers from one place to another. For instance, an eight-bit number normally requires eight connections (each one representing one bit). But the same number can be transferred serially via a single connection.

For example, Figure 5.18 shows the voltage waveform for an eight-bit number sent serially. Each bit is transferred within a given time slot, requiring eight time slots to transfer eight bits. To interpret the waveform correctly, the length of the time slot, the logic polarity (positive or negative logic), and the order of the bits must all be known.

If the serial data is sent repetitively, either an analog or digital scope can display the waveform. More likely, the serial data is sent randomly in bursts, perhaps with a different set of bits in each burst. In this case, a digital scope is required to capture these events that occur only once or infrequently. The scope must have sufficient sample rate and memory depth to capture the required serial data in one acquisition.

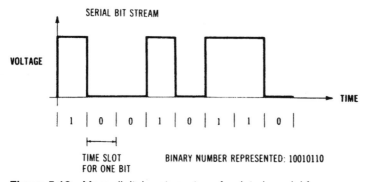

Figure 5.18 Many digital systems transfer data in serial form.

5.5.4 Digital Counter

Another example of a digital circuit with complex waveforms is the digital counter. A 4-bit binary counter is shown in Figure 5.19. The binary outputs (Q0 through Q3) represent a 4-bit binary number. On each rising edge of the clock input, the binary number increments by one. Thus, the counter counts the number of rising edges of the clock and the binary counting pattern is shown in the figure.

The timing waveforms of the digital outputs are plotted in Figure 5.20, assuming that the counter is originally in the 0000 state. Notice that all changes in the outputs occur on the

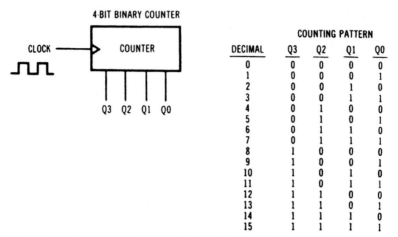

Figure 5.19 A 4-bit binary counter counts the number of rising edges on the clock input.

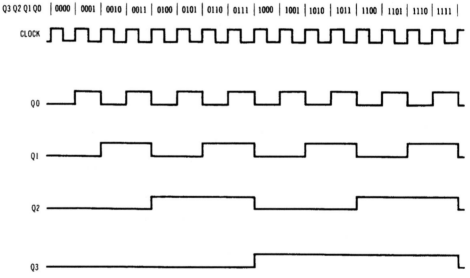

Figure 5.20 The timing waveforms for a 4-bit digital counter, corresponding to the logic states shown in Figure 5.19.

rising edge of the clock. It's also worth pointing out that the frequency of the Q0 output is one-half the clock frequency (divide-by-2). Similarly, the frequency of the Q1 output is one-fourth the clock frequency (divide-by-4), Q2 is one-eighth the clock frequency (divide-by-8), and Q3 is one-sixteenth the clock frequency (divide-by-16). So a binary counter can be used for dividing down a clock to generate submultiple frequencies as well as counting clock pulses.

Suppose we need to verify the proper operation of a 4-bit counter. How would we connect an oscilloscope to the counter and what would be the best signal to trigger on? Figure 5.20 shows the five logic signals associated with the counter. First, we might check that the clock signal is present by viewing it with the scope. Then, we should check the timing relationship of Q0, Q1, Q2, and Q3. Assuming that we only have simple edge trigger available, one might be tempted to trigger the scope on the rising edge of the clock, since it is the counter's main timing signal. There are 16 rising edges of the clock in Figure 5.20, which means that there are 16 possible trigger events. Sometimes the scope will trigger on the first edge, sometimes on the second edge, and so forth. The oscilloscope display of Q0 through Q3 will not be stable as the various trigger events occur. Since the waveforms will move around, measuring the timing relationships will be difficult. We need to stabilize the scope display using one of the following techniques:

1. *Trigger holdoff.* With the clock used as the trigger source, adjust the trigger holdoff control to stabilize the waveform display. Holdoff can be adjusted via trial-and-error or set so that it equals 16 clock periods on the counter. When the right amount of holdoff is selected, only one trigger per counter cycle will be processed—the other triggers will be ignored.

2. *Trigger on lowest frequency signal.* Another solution is to choose a different trigger source. Note that Q3, the lowest frequency in the system, has only one rising edge in Figure 5.20. Of course, Q3 will have other trigger events as the signal repeats beyond the edge of the figure, but each time Q3 goes high the other signals will have exactly the same logic state. Therefore, triggering on the lowest frequency signal causes the higher frequency signals to be stable on the oscilloscope display.

3. *Pattern trigger.* As discussed in Chapter 4, pattern triggers can be used to trigger on the complex waveforms associated with the 4-bit counter. Using a 4-channel scope, we can set the trigger condition as one of the counter states (Q3, Q2, Q1, Q0) shown in Figure 5.19. For example, we could set the trigger pattern as all channels low (L L L L) such that the scope would trigger when the counter hit zero. Actually, we could use any of the 16 counter states to get a stable trigger.

4. *Single acquisition.* Another approach to verifying the counter operation is to forget triggering completely and capture the waveforms with one acquisition using a digital scope. As long as the scope has sufficient sample rate and memory depth to capture a complete cycle of the circuit operation, we can make the measurement without worrying about a stable trigger. However, we may be limited by the number of scope channels. A 4-channel scope is desirable here since it can display all four Q outputs simultaneously, while a 2-channel scope would force the user to move the scope probes from signal to signal. Using the untriggered capture approach while moving from signal to signal on the counter can be very confusing, as the time axis will be different for each measurement.

All four of these approaches can result in a stable display of the counter's waveforms. Once a stable display is obtained, we can use the scope to verify that the timing relationships of the counter are correct.

The preceding discussion used the example of a 4-bit counter, which is a simple, almost trivial circuit in today's complex digital world. However, the basic principles that have been explored can be applied to more complex digital circuits.

5.5.5   D-type Flip-Flop

A D-Type flip-flop has an input (D), an output (Q), and a clock input (Figure 5.21). The output remains in the same state until a rising edge of the clock is encountered. At that time, Q takes on the logic level present at the D input. Q will remain in the state until the next rising edge of the clock. The net result is that the Q output tends to track the D input, but only changes on a rising clock edge.

Some typical voltage waveforms are shown in Figure 5.21a. The key to understanding the circuit operation and the waveforms is to focus in on the rising clock edges. Only then can the output change state. On the first rising edge of the clock, Q changes from 0 (its previous state) to 1 (the logic level currently present at the D input). Q stays in this state until the second rising clock edge, at which time D is 0, so Q becomes 0. Notice that although the logic level at the D input changes several times, these changes are transferred to the Q output only upon a rising edge of the clock. Also, no action occurs on the falling edge of the clock.

For the flip-flop to operate correctly, the D input must be stable at a valid logic level for some time before the rising clock edge, called the *setup time*. There may also be a requirement that the D input remains stable for a short time after the clock edge, which is called the *hold time*. On many flip-flops, there is no hold time requirement, which is to say that the required hold time is zero.

A scope can be used to check for setup and hold time violations by viewing the clock, D input, and Q output signals while triggering on the rising clock edge (Figure 5.21b). To meet the setup time requirement, the D input must be above the high logic threshold or below the low logic threshold during the setup time period in front of the clock edge. The hold time requirement means that the D input must remain a valid logic high or low after the rising clock edge until the hold time expires. Most circuits will have instances where D is high at the clock edge and instances where D is low at the clock edge. The oscilloscope display will show both high and low states, similar to the upper waveform in Figure 5.21b.

A setup or hold time violation might occur infrequently, perhaps due to a particular sequence of logic states or timing variations. The scope user watching the waveform display might miss such an occasional violation. Recall from Chapter 4 that *infinite persistence* causes all acquisitions of the waveform to be accumulated onscreen. A powerful technique for finding infrequent events is to turn on infinite persistence and let the scope run for a long time. The user can check back later to see if any of the accumulated waveforms had a setup or hold time violation.

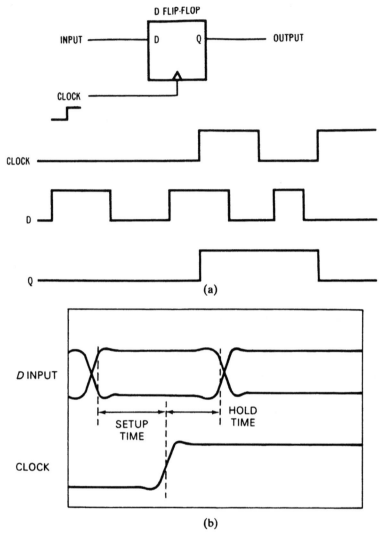

Figure 5.21 (a) The Q output of a D flip-flop changes to the same logic level as the D input on each rising edge of the clock. (b) The setup time and hold time for the D flip-flop can be measured using a scope.

Triggering directly on a fault is a reliable way of capturing any kind of infrequent failure. Scopes equipped with setup time trigger and hold time trigger (as discussed in Chapter 4) can trigger directly on a setup or hold time violation, without the user having to worry about whether an event was missed.

5.5.6   Digital Measurement Hints

It should be apparent from the above examples that digital circuit operation can produce some very complex waveforms and some very difficult measurement challenges. Based on these examples, we can propose some general principles for measuring digital signals with a scope:

- Use a scope with four channels or a mixed signal oscilloscope. Digital systems are complex and have many signals that need to be viewed to understand the circuit operation.
- Use trigger holdoff to help stabilize the display of complicated waveforms.
- Trigger on the lowest frequency signal in the system when viewing multiple digital signals.
- Use advanced triggering features (pattern trigger, setup time trigger, hold time trigger) to stabilize the waveform display and find elusive problems.
- Use single acquisition measurements with a digital scope to sidestep triggering problems.
- Use of infinite persistence to accumulate a large number of waveform acquisitions, which is helpful in finding infrequent events.

5.6   Frequency Response Measurement

Earlier in the chapter, gain and phase measurements were discussed as applied to a single frequency. A single-frequency sine wave was connected to the input of a circuit and the gain through the circuit as well as the phase of the output signal (relative to the input) were measured. This describes the behavior of that circuit at that particular frequency, but it is often desirable to characterize the circuit performance over a wide range of frequencies. The gain measured over a range of frequencies is called the *frequency response* of the circuit.

5.6.1   Combined Single-Frequency Measurements

One way to measure the frequency response of a circuit is to perform multiple single-frequency gain measurements and plot them as gain versus frequency. For example, Figure 5.22 shows a simple RC low-pass filter being driven by a sine wave source. The oscilloscope is connected such that it measures the input voltage and output voltage of the circuit. The signal source is stepped through the range of frequencies with voltage readings recorded at each frequency. The resulting voltage measurements at the input and output are tabulated in Table 5.1. One might be tempted to assume that V_{IN} will always be constant, but this may not be true. V_{IN} can change due to source flatness and/or loading effects changing with frequency.

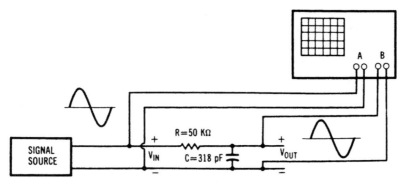

Figure 5.22 A frequency response measurement on an RC low-pass filter.

Table 5.1 Measured Values for RC Circuit

Frequency (Hz)	V_{IN} (volts)	V_{OUT} (volts)	Gain	Gain (dB)
2000	0.200	0.196	0.980	−0.18
4000	0.200	0.186	0.930	−0.63
6000	0.200	0.172	0.860	−1.31
8000	0.200	0.156	0.780	−2.16
10000	0.200	0.142	0.710	−2.97
12000	0.200	0.128	0.640	−3.88
14000	0.200	0.116	0.580	−4.73
16000	0.200	0.106	0.530	−5.51
18000	0.200	0.098	0.490	−6.20
20000	0.200	0.090	0.450	−6.94
30000	0.200	0.064	0.320	−9.90
40000	0.200	0.048	0.240	−12.40
50000	0.200	0.040	0.200	−13.98
60000	0.200	0.032	0.160	−15.92
70000	0.200	0.028	0.140	−17.08
80000	0.200	0.024	0.120	−18.42
90000	0.200	0.022	0.110	−19.17
100000	0.200	0.020	0.100	−20.00

The gain for each frequency is calculated by dividing the output voltage by the input voltage at that frequency. The gain versus frequency plot is shown in Figure 5.23. The gain can also be expressed in decibels, as shown in Table 5.1, and the gain in decibels versus *log-*

arithmic frequency is shown in Figure 5.24. The dB scale is inherently logarithmic, which results in a log versus log plot, which is a common way to show frequency response information. This logarithmic scale has the benefit of showing a wide range of gain and frequency on a compact plot. Note that the frequency scale of Figure 5.24 easily accommodates several decades of frequency, while the linear scale used in Figure 5.23 does not.

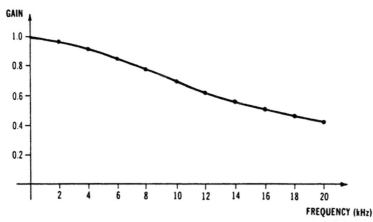

Figure 5.23   The frequency response of the RC filter plotted as linear gain versus linear frequency.

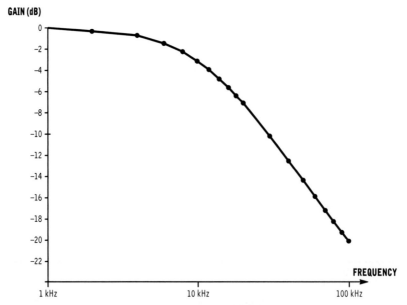

Figure 5.24   The frequency response of the RC filter plotted as gain in decibels versus log frequency.

5.6.2   Swept-Frequency Response

Although the previously described point-by-point method is valid and produces accurate results, it is somewhat time consuming. Another method of measuring frequency response involves using a sweep generator to speed up the measurement. The swept sine wave of the sweep generator is connected to the input of the circuit under test (Figure 5.25). The output of the circuit under test is connected to the vertical channel of an oscilloscope, operating in the X-Y mode. The sweep voltage of the sweep generator drives the horizontal axis of the scope. As the sweep generator sweeps in frequency, the sweep voltage of the generator ramps up (in proportion to frequency), causing the output of the circuit under test to be plotted across the scope display. In this manner, the entire frequency response of the circuit is quickly displayed on the scope (Figure 5.26).

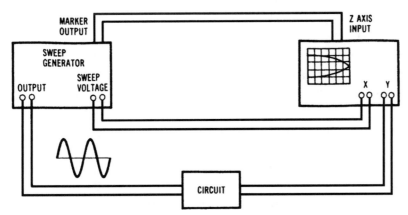

Figure 5.25   A sweep generator and a scope operating in X-Y mode can be used to plot the frequency response of the circuit automatically.

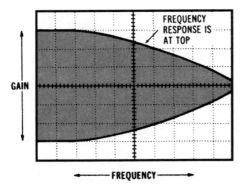

Figure 5.26   The oscilloscope display of the swept-frequency response. The vertical axis is V_{OUT} and the horizontal axis is the frequency range swept by the generator.

This method relies on the output of the sweep generator being constant with frequency. Imperfections in the flatness of the generator will cause an error in the frequency response. The drive capability of the generator is important since the load that the circuit places on the generator will usually vary over frequency. The generator must be relatively insensitive to these load changes or additional error will be introduced. For both of these reasons it is a good idea to check with the scope to make sure that V_{IN} is constant as the generator sweeps.

The oscilloscope is displaying V_{OUT}, and not the gain (V_{OUT}/V_{IN}). If V_{IN} was set up to be a convenient value (such as 1 volt), then the display can be interpreted directly as gain. Otherwise, some mental arithmetic is necessary to convert the V_{OUT} display to the actual gain value.

The sweep generator should not be swept too fast, since the circuit under test needs time to respond. This is particularly important in circuits that have abrupt changes in gain as the frequency varies. The sweep rate for the generator is usually set experimentally by reducing the sweep rate until the frequency response no longer changes with each change in sweep rate. The sweep generator may be swept in either a linear or logarithmic manner depending on the desired type of frequency axis. The oscilloscope vertical axis is, of course, always linear.

Many sweep generators supply a *marker* output signal, which pulses when a particular frequency (or frequencies) is present at the generator output. This signal can be connected to the Z axis input of the oscilloscope so that the location of the marker frequency can be accurately identified on the scope display. The sweep generator will pulse the marker output, causing a change in intensity on the oscilloscope display at precisely the marker frequency. Exactly how the intensity changes (whether it gets brighter or dimmer) will depend on the polarity of the marker signal as well as the polarity of the Z axis input.

Another type of marker function is sometimes provided, which pulses the output voltage slightly at the marker frequency. This causes a "blip" to appear on the display at the marker frequency. The Z-axis input is not used in this case.

5.7   Square Wave Test

Sine waves are the predominant signal for characterizing the frequency response of analog circuits, but another form of frequency response test can be performed using a square wave. The square wave test is a quick, qualitative method for checking a circuit for flat frequency response.

A classic use of the square wave test is in high-fidelity audio equipment. The standard test for audio is to use a 1 kHz square wave to check for overall flat response in the system. Another use of square wave testing is the compensation of attenuating probes already discussed in Chapter 4.

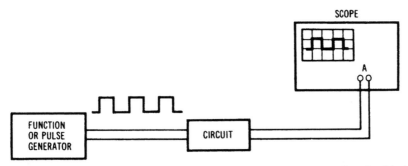

Figure 5.27   The square wave test is a good qualitative test of a circuit's frequency response.

A square wave is applied to the input of the circuit being tested and the output of the circuit is monitored on an oscilloscope (Figure 5.27). As discussed in Chapter 1, the square wave is very rich in harmonics, extending out to many times its fundamental frequency. The relative amplitude of each of these harmonics must remain unchanged for the output to be a square wave. If the circuit under test has gain or loss, the amplitude of each harmonic is increased by the gain or loss, but their amplitude relative to each other should remain the same if the frequency response is flat. In addition, each harmonic has a particular phase relationship with the fundamental frequency that must be maintained or the shape of the output waveform will be distorted. The circuit could have a perfectly flat amplitude response, but not pass a square wave correctly due to phase distortion.

This type of test requires a high-quality waveform at the input. Depending on the output characteristics of the generator, the circuit under test may load the generator enough to cause distortion, so it is a good idea to monitor the input waveform of the oscilloscope.

Although the square wave test does not result in a frequency response plot, it does test a circuit quickly, with good qualitative results. Some typical output waveforms encountered in square wave testing and their causes are shown in Figure 5.28. Phase shift at low or high frequencies can cause a tilt to one side or the other of the square wave. It is difficult to predict the effects of attenuation at either high or low frequencies, as it depends greatly on the exact shape of the frequency response. However, a few typical examples are shown. Some circuits will exhibit a noticeable degradation in the rise time of the output square wave, referred to as *slew-rate limiting*. Due to the nature of the square wave test, it is more effective at determining whether a problem exists than identifying the exact problem.

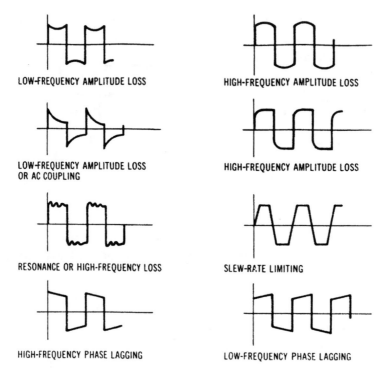

LOW-FREQUENCY AMPLITUDE LOSS

HIGH-FREQUENCY AMPLITUDE LOSS

LOW-FREQUENCY AMPLITUDE LOSS
OR AC COUPLING

HIGH-FREQUENCY AMPLITUDE LOSS

RESONANCE OR HIGH-FREQUENCY LOSS

SLEW-RATE LIMITING

HIGH-FREQUENCY PHASE LAGGING

LOW-FREQUENCY PHASE LAGGING

Figure 5.28   Some examples of waveforms resulting from the square wave test.

5.8   Linearity Measurement

It is often desirable to measure the DC voltage out of a circuit compared to the DC voltage in. This can be done using the X-Y mode of the oscilloscope, as shown in Figure 5.29. A

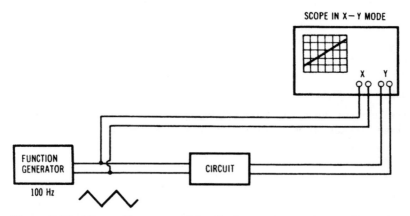

Figure 5.29   The oscilloscope and function generator can be used to measure the DC linearity of a circuit.

slowly varying DC voltage is applied to the input of the circuit as well as the X axis of the scope. A convenient method of obtaining the varying DC voltage is to use a triangle wave with a very low frequency. The output of the circuit is connected to the Y axis of the scope, resulting in the output voltage being plotted versus the input voltage. The frequency of the generator is chosen fast enough so that the display updates quickly enough, but slow enough that the operation of the circuit is not affected. (Remember, a DC voltage is being simulated.)

5.8.1   Clipping Circuit

Consider the clipping circuit shown in Figure 5.30. When V_{IN} is a positive voltage less than 5 volts, the zener diode acts like an open circuit and V_{OUT} is equal to V_{IN}. When V_{IN} becomes greater than 5 volts (the zener voltage), then the diode turns on and V_{OUT} is limited to 5 volts, no matter how large V_{IN} gets. If V_{IN} becomes less than zero, then the diode turns on in the other direction and limits V_{OUT} to about –0.6 volts, depending on the diode. The effect of the circuit is to limit the output voltage to between about –0.6 volts and +5 volts.

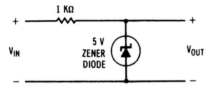

Figure 5.30   This clipping circuit limits the output voltage to between –0.6 volts and 5 volts.

Using the DC linearity measurement technique described, the V_{OUT} versus V_{IN} linearity plot that is obtained is shown in Figure 5.31. The sloped portion of the trace corresponds to the region where V_{OUT} equals V_{IN}. To the left, the diode clamps the voltage at –0.6 volts while on the right the voltage is limited at +5 volts.

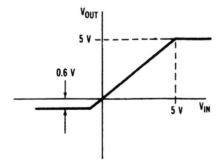

Figure 5.31   The linearity plot (V_{OUT} versus V_{IN}) for the clipping circuit.

5.8.2   Amplifier

A properly functioning amplifier produces a waveform at the output that is an exact replica of the input, but amplified by the gain of the amplifier. This is expressed mathematically as

$$V_{OUT} = G \cdot V_{IN}$$

Thus, the output waveform is the same as the input waveform, except scaled by the gain of the amplifier, G. For increasing input voltage, the output increases accordingly. However, in a practical amplifier, there is a point at which the output voltage saturates and will not go any higher. If the input to the amplifier is a sine wave, the output waveform will be clipped, as shown in Figure 5.32.

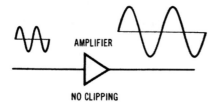

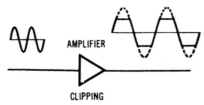

Figure 5.32   Clipping occurs at the output of an amplifier when the amplifier cannot produce the peak voltage of the waveform.

The V_{OUT} versus V_{IN} linearity plot of the amplifier can also be used to measure this phenomenon (Figure 5.33a). The display is a straight line whose slope is the gain of the amplifier. Ideally, the straight line of the display would extend indefinitely. That is, the amplifier would be capable of amplifying any input voltage, no matter how large. In reality, the amplifier will clip at some point, usually as the peak voltage approaches the amplifier's power supply voltage, as shown in Figure 5.33b. The straight line is still present, but flattens out at the point of limiting. Typically, the trace does not break sharply, but instead is rounded off.

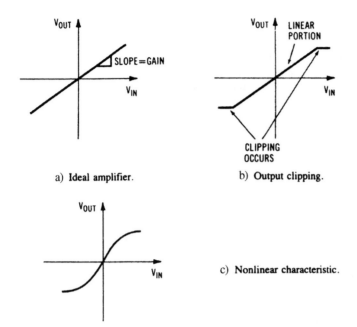

Figure 5.33 (a) The V_{OUT} versus V_{IN} plot for an ideal amplifier is a straight line. (b) Amplifier clipping is shown as a flattening of the linearity plot. (c) A curved plot indicates nonlinear operation in the amplifier.

Since the V_{OUT} versus V_{IN} characteristic is a straight line, this type of circuit operation is called *linear*. If the plot is not a straight line, then the characteristic is termed *nonlinear* (Figure 5.33c). Nonlinear amplifier operation causes distortion of the output signal, usually in the form of harmonics.

5.9   Curve Tracer Measurement Technique

The curve tracer measurement technique is used to measure the characteristics of components such as diodes and resistors. This technique uses the oscilloscope in X-Y mode to display the current through a component versus the voltage across the component. This display is referred to as the I-V (current–voltage) characteristic of the component.

There are several ways to make this measurement. Figure 5.34a shows a function generator driving the curve tracer circuit. V_2 is the voltage across the device being measured and V_1 is the voltage across the resistor. By Ohm's Law, V_1 is proportional to the current through the resistor, which is also the current through the component under test. So if V_1 could be displayed versus V_2, the I-V characteristic of the component would be shown. If the scope has floating inputs, then this can be done with no problem.

Unfortunately, most scopes have grounded inputs, which results in the situation shown in Figure 5.34b. Both sides of the component under test are grounded, resulting in a short cir-

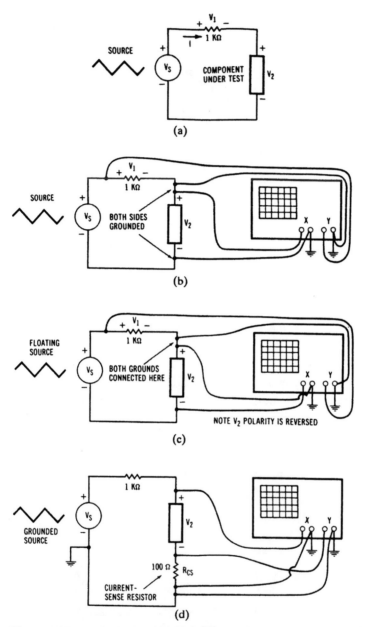

Figure 5.34 Methods for displaying the I-V characteristics of a component. (a) The approach suitable for use with floating scope inputs. (b) Grounding problems when using typical scope with grounded inputs. (c) An approach for use with a floating source. (d) Use of a current-sense resistor.

cuit across it. The situation is compounded even further if the source has a grounded output. If the source is floating, then the circuit in Figure 5.34c can be used. Note that the circuit ends up being grounded at only one point, which is where both scope input grounds are connected. Something has changed, though. The horizontal axis of the scope will be $-V_2$ (instead of $+V_2$) due to the reversal of the horizontal input leads. This can be compensated for if the scope has an "Invert Channel" switch, otherwise the I-V curve will appear backward on the display (left half of the display swapped with the right half). This may be acceptable if the user is willing to mentally convert it back.

If both the scope and the source are grounded, then another technique must be used. A current sense resistor is placed in series with the component under test. The vertical scope input then uses the voltage across this resistor to measure the current. The voltage across the current-sense resistor will introduce a small error in the measurement of V_2, but as long as the resistor is kept small the error will be acceptable. The resistor cannot be made too small since, for a given current being measured, the voltage will decrease with smaller resistance. The sensitivity of the scope will ultimately determine how small the current sense resistor can be made.

5.10   Diode I-V Characteristic

Figure 5.35 shows an oscilloscope set up to measure the I-V characteristics of a diode, using the current sense method. The source (usually a function generator) is set to produce a low-frequency triangle wave, although a sine wave will also work. The triangle wave acts as an automatically varying DC voltage, causing the voltage across the diode to also change. At the same time, the voltage across and the current through the diode are measured. The amplitude and frequency of the function generator can be set experimentally, but a zero-to-peak voltage of 5 volts and a frequency of 30 Hz is a good starting point.

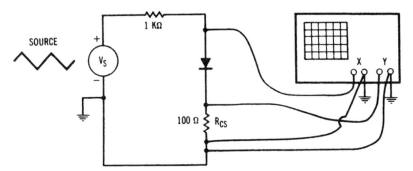

Figure 5.35   The I-V characteristic of a diode can be measured using the curve tracer circuit with current-sense resistor.

The resulting I-V characteristic for a diode is shown in Figure 5.36. The horizontal scale can be determined directly from the volts/div setting. The vertical scale must take into account the value of the current sense resistor.

$$amps \ per \ div = (volts \ per \ div)/R$$

where R is the value of the current sense resistor.

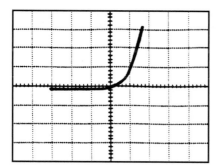

Figure 5.36   The I-V curve of a semiconductor diode.

The resulting curve is the classical behavior of the semiconductor diode. For voltages greater than zero (right half of the display), the current quickly increases. For voltages less than zero (left half of the display), the current is essentially zero. So the diode conducts in the forward direction, but acts like an open circuit in the reverse direction.

If the test is not set up properly, the scope display may show two separate traces instead of the single curve. One trace is drawn as the voltage increases and the other is drawn on the decreasing portion of the triangle wave. They may be slightly different due to either capacitive effects or heating of the component being tested. Reducing either the frequency or the amplitude of the triangle wave will cause the two traces to converge into one single trace.

5.11   Resistor I-V Characteristic

The curve tracer circuit can be used to measure unknown resistors. The resulting I-V curve is shown in Figure 5.37. The current sense resistor, R_{CS}, should be chosen to be at least a factor of 10 smaller than the unknown resistance. The voltages shown in Figure 5.37 (ΔX and ΔY) should be determined, taking into account the volts/div settings. The resistance value is determined by calculating 1/slope of the line, using the equation shown below.

$$R = 1/slope$$

$$R = \frac{\Delta X \, R_{CS}}{\Delta Y}$$

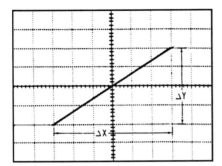

Figure 5.37 The I-V curve of a resistor is a straight line with R = 1/slope.

Measuring resistor values is probably done more efficiently by using the resistance function of an accurate digital multimeter. However, the curve tracer method shown here provides more than just the resistance value. The I-V plot shows us the behavior of the resistor as a function of voltage, including any nonlinearities or breakdown effects that may be present in the device.

5.12 Amplitude Modulation Measurement

As presented in Chapter 1, an amplitude modulated signal is represented by the equation

$$v(t) = A_C[1 + am(t)]\cos(2\pi f_c t)$$

where

A_c = the signal amplitude

a = modulation index ($0 \le a \le 1$)

m(t) = normalized modulating signal (maximum value is 1)

f_c = carrier frequency

The modulation percentage of an amplitude modulated signal can be measured in the time domain with an oscilloscope. The timebase of the scope is set up to view the modulating frequency (not the carrier frequency). As shown in Figure 5.38, the modulating frequency shows up in the envelope of the carrier. The modulation percentage is determined by noting the maximum and minimum of the envelope, V_{max} and V_{min}. The maximum of the envelope voltage occurs when the modulating signal is at its most positive value, +1, and the minimum occurs when the modulating signal is –1.

$$V_{max} = 1 + a$$

$$V_{min} = 1 - a$$

Solving for the modulation index,

$$a = \frac{V_{max} - V_{min}}{V_{max} + V_{min}}$$

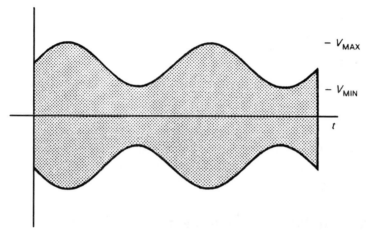

Figure 5.38 An amplitude modulation measurement is made using the minimum and maximum of the envelope of the signal.

5.13 Power Measurement

Scopes are most often used for voltage measurements, but through the use of current probes they can also measure current waveforms. Figure 5.39 shows a measurement of the input to a switching power supply. The larger waveform is the sinusoidal voltage from the power line and the smaller waveform is the current being delivered to the supply. Note that while the voltage waveform is roughly a sine wave,[4] the current waveform consists of rounded pulses that occur near the voltage peaks. This is typical of a modern power supply.

Since we are dealing with waveforms that are from the power line, they are inherently synchronized with the power-line frequency. Therefore, we can use *line trigger* as a convenient way of triggering the scope without having to adjust the trigger level or other trigger features.

The automatic measurements on the scope display indicate that the frequency of the line voltage is approximately 60 Hz. The voltage waveform is 325 volts peak-to-peak or 118 volts RMS. The current probe has a scale factor of 10 mV/A, meaning that a 1-amp current

---

4.   The power-line voltage is a distorted sine wave, indicating the quality of electrical power at the
       author's home office.

through the probe produces 10 mV on the scope display. The scope display indicates 10 mV/div (channel 2) for the current waveform, so taking the probe factor into account means that the channel is operating at 1 A/div.

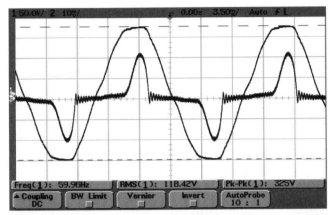

Figure 5.39 Voltage and current measurements of a switching power supply.

Using the math function of a digital scope allows us to multiply the voltage and the current to produce instantaneous power.

$$p(t) = v(t) \cdot i(t)$$

The plot of power versus time for the voltage and current from Figure 5.39 is shown in Figure 5.40. The center of the screen (vertically) is zero, so the power waveform starts at zero and rises to a peak, pulsing positively at twice the frequency of the original 60-Hertz

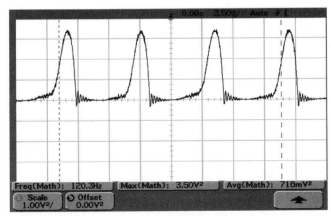

Figure 5.40 This is the instantaneous power waveform derived by multiplying the voltage and current waveforms from Figure 5.39.

line frequency. The scope display shows that the frequency of the power waveform is 120 Hertz and the maximum of the waveform is 3.5 V^2. The probe factor for the current probe is 10 mV/A, so taking the reciprocal gives us 100 A/V. To get the maximum power, we multiply 3.5 V^2 × 100 A/V = 350 watts. However, the average power in the waveform is much less, found by taking the average of the waveform shown on the display as 718 mV^2 and multiplying it by 100 A/V to get 71.8 watts.

The output of this switching supply is nominally 12 volts DC. Figure 5.41 shows the ripple and noise on the DC output. (Note that the volts/division is set at a very sensitive 10 mV/div.) Again, we can use the line trigger feature on the scope since the ripple is associated with the line frequency. The large amount of high-frequency noise on the waveform is switching noise from the power supply and digital noise from other circuits in the system.

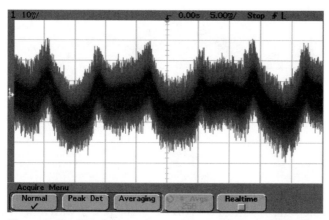

Figure 5.41   This is the noise and ripple on the DC output of the power supply.

It is difficult to see the low frequency (60 Hz or 120 Hz) ripple on the power supply due to the high-frequency noise that is on the waveform. The scope measurement already has the bandwidth limit enabled on the channel, but much of the noise still gets through. To get a better look at the ripple on the power supply, we can remove the high-frequency noise by averaging the waveform while triggering on the line frequency. Any signals that are not synchronous with the trigger will tend to average away, leaving only trigger-synchronous signals. In this case, the noise is removed while the ripple related to the line-frequency remains. Figure 5.42 shows the waveform that is produced by averaging out the noise. Now we can see that there is 14.5 mV peak-to-peak of ripple on the supply output.

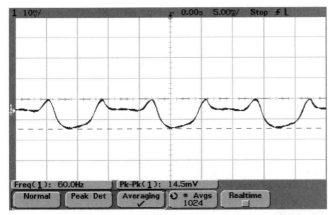

Figure 5.42 Averaging while using line trigger removes the high-frequency noise and exposes the ripple on the power supply output.

5.14   FFT Measurements

The Fast Fourier Transform (FFT) is a very powerful mathematical algorithm that transforms the time domain view of a waveform into the frequency domain.[5] The FFT is a natural feature for a scope since it can convert the time domain data inherent in a scope to a frequency domain view that exposes the frequency content of the waveform.

The FFT is a specific implementation of the Discrete Fourier Transform, which is the discrete (or digitized) version of the Fourier Transform.[6] Entire books have been devoted to explaining the theory and application of the FFT, so we won't try to recreate that work here.[7]

The basic concept of the FFT algorithm is shown in Figure 5.43. The input is a set of equally spaced time-domain samples of the waveform, sampled at a rate f_s. The time period between samples is

$$T_s = 1/f_s$$

For N time domain samples, there are usually N/2 frequency domain samples, which represent frequencies from 0 to $f_s/2$. The frequency domain data will be spaced by f_s/N. Recall that the scope sample rate can vary with the timebase setting, so the frequencies represented by the FFT may also vary with timebase setting, due to the dependency on sample rate. The output of the FFT is in terms of linear voltage but it is common to display the FFT output in terms of decibels.

5.   J.W. Cooley and O.W. Tukey developed the FFT in 1965.
6.   The Fourier Transform is named after the mathematician John Baptiste Joseph Fourier (1768–1830).
7.   See Brigham, 1997 or Ramirez, 1985.

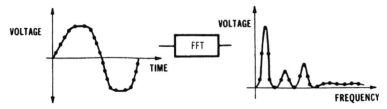

Figure 5.43   The FFT algorithm converts the time domain samples into the frequency domain representation.

Figure 5.44 shows a 20-kHz sine wave (upper trace) and the FFT of the sine wave (lower trace). The sample rate of this measurement is 1 MSa/sec producing an FFT frequency span that goes from 0 to 500 kHz (which is $f_s/2$). The FFT shows a large spectral line near the left side of the display, which represents the 20-kHz energy of the sine wave. The rest of the FFT plot is the noise floor of the measurement.

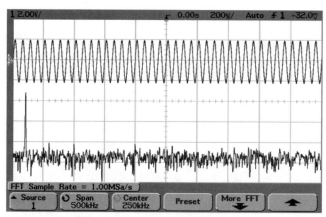

Figure 5.44   The upper trace is a 20-kHz sine wave and the lower trace is the FFT display of the same waveform.

Using the same measurement configuration shown in Figure 5.44, increasing the frequency of the sine wave causes the cycles of the sine wave to compress together and the FFT spectral line moves to the right. Figure 5.45 shows the measurement result with the sinewave frequency at 250 kHz. The sine wave is now shown as a solid band since the period of the waveform is very small compared to the timebase setting of the scope. At the same time, the spectral line shows up at the middle of the frequency span in the FFT plot.

Figures 5.44 and 5.45 illustrate a fundamental characteristic of the FFT algorithm. When a signal is scaled in time for convenient viewing in the time domain, the corresponding spectral line shows up near the left of the FFT plot. When the timebase is adjusted to move the spectral line to the middle of the FFT plot, the time domain waveform is very compressed.

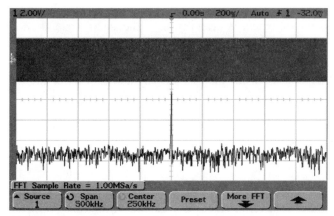

Figure 5.45   The frequency of the sine wave is increased to 250 kHz which causes the upper trace to be a solid band. The FFT plot now shows the spectral line at 250 kHz (center of the frequency span).

Some scopes will allow the user to rescale the FFT frequency span without changing the timebase setting so that both the time domain view and frequency domain view can be optimized.

5.14.1 Aliasing

As implemented in most scopes, the FFT algorithm is implemented as a math function that operates on the scope waveform samples, without regard to the Sampling Theorem. As the timebase setting and sample rate are changed, the sample rate may fall below the Nyquist Rate and aliasing can occur.

Aliasing in the time domain was already discussed in Chapter 4. Aliasing in the frequency domain can be observed in Figure 5.46. This is the same measurement setup as Fig-

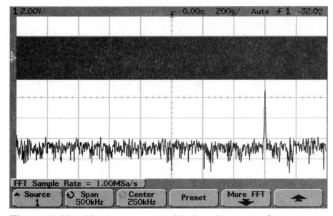

Figure 5.46   Aliasing occurs with the sine wave frequency at 600 kHz. The spectral line "folds back" and appears at 400 kHz in the FFT display.

ures 5.44 and 5.45 but with the sine wave frequency now set at 600 kHz. Recall that the upper frequency limit of this measurement is determined by the sample rate (1 MSa/sec), which produces an upper frequency limit of $f_s/2$, or 500 kHz. The frequency of the 600-kHz sine wave exceeds $f_s/2$ and should fall outside of the displayed frequency span. Instead, due to aliasing, the frequency "folds back" inside the FFT display and appears at 400 kHz. So instead of being 100 kHz outside of the frequency range, it folds back inside of the FFT range and is displayed at $f_s/2 - 100$ kHz, or 400 kHz.

5.14.2 Spectral Leakage

One limitation of the FFT is that it operates only on one piece of the waveform, finite in length, as acquired by the scope. The FFT algorithm has the effect of extending this time record for all time by replicating the waveform that was captured.[8] Unfortunately, in the general case, this replication introduces a transient into the time record, which shows up as smearing in the frequency domain called *spectral leakage*. Figure 5.47 shows a signal that should be an infinitely thin spectral line that has significant side lobes due to spectral leakage.

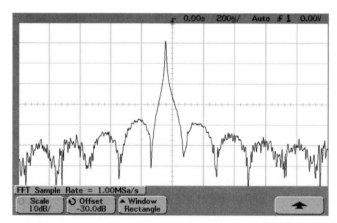

Figure 5.47 This measurement of a pure spectral line using a *rectangular* time window exhibits spectral leakage.

The solution to spectral leakage is the use of a windowing function that applies a weighting function to the time domain record. Figure 5.48 shows the same signal as Figure 5.47, measured using a *Hanning* (also known as *Hann*) window on the time domain data. The spectral leakage is virtually eliminated. Figure 5.49 shows the same signal measured with a *flattop* window and also gives a leakage-free measurement. The measurement in Figure 5.47

---

8.   For a more complete explanation of spectral leakage and windows, see Witte, 2001.

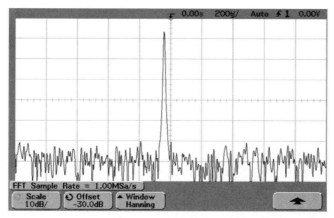

Figure 5.48 This measurement of a pure spectral line using a *Hanning* time window eliminates spectral leakage.

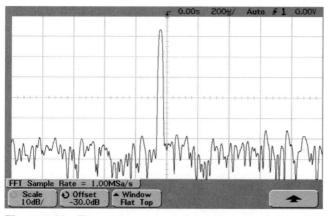

Figure 5.49 This measurement of a pure spectral line using a *flattop* time window eliminates spectral leakage.

is said to use a *rectangular* window where all time domain data is equally weighted. This is the same as saying the data is not windowed at all.

A careful examination of Figure 5.48 and Figure 5.49 reveals that the shape of the spectral line depends on the window being used. The Hanning window produces spectral lines that are more thinner and more peaked, while the flattop window produces fatter spectral lines. What is less obvious is that the more rounded shape of the flattop window defines the top of the spectral line and will result in better amplitude accuracy than the Hanning window. However, this amplitude accuracy comes at the expense of frequency resolution due to the wider spectral line.

There are many other time domain windows that can be applied to the FFT but they all have the general tradeoff between frequency resolution, amplitude accuracy, and suppression of spectral leakage. For most applications, any of the common windows will be acceptable (excluding the rectangular window) and the scope will usually default to a reasonable choice. For more critical applications, the scope user will need to select the window that optimizes the parameters that are most important.

5.14.3 Harmonic Distortion

The 100-kHz waveform shown in Figure 5.50 has some noticeable distortion near the peaks of the waveform. A pure sine wave should result in one spectral line in the frequency domain. However, this waveform will exhibit *harmonic distortion,* which is visible in the frequency domain. Figure 5.51 is the FFT of this same waveform. The left-most spectral line is the 100-kHz fundamental of the waveform. The next spectral line to the right is at

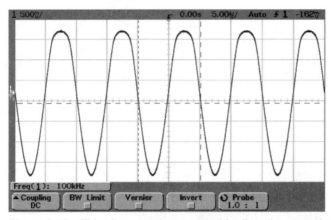

Figure 5.50   This sine wave has noticeable distortion near the peaks of the waveform.

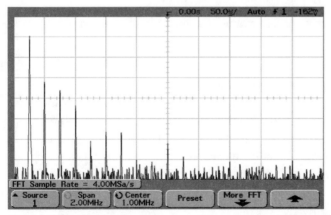

Figure 5.51   The harmonic content of the waveform in Figure 5.50 is shown here using the FFT function. The vertical scale is 10 dB/div and the frequency span is 0 to 2 MHz.

twice the fundamental frequency, so it is the second harmonic. The remaining spectral lines are higher-order harmonics. The second harmonic is the largest harmonic and is slightly more than 20 dB below the fundamental.

5.14.4 Amplitude Modulation

The amplitude-modulated signal in Figure 5.52 has a carrier frequency of 100 kHz and a modulation frequency of 10 kHz. The carrier is 50% modulated so a, the modulation index, is equal to 0.5.

Recall from Chapter 1 that an AM signal is described by the equation

$$v(t) = A_c \cos(2\pi f_c t) + \frac{aA_c}{2} \cos[2\pi(f_c + f_m)t] + \frac{aA_c}{2} \cos[2\pi(f_c + f_m)t]$$

In the frequency domain, this equation corresponds to a carrier frequency, f_c (with amplitude A_c), an upper sideband, $f_c + f_m$ (with amplitude $aA_c/2$) and a lower sideband, $f_c - f_m$ (with amplitude $aA_c/2$). The ratio of the sideband amplitude to the carrier amplitude is expressed in decibels as:

$$AM\ Sideband\ Level(dBc) = 20 \log \left[\frac{aA_c/2}{A_c} \right] = 20 \log(a/2)$$

For this case, $a = 0.5$, so the AM Sideband Level (dBc) = –12 dBc. Figure 5.53 shows the FFT display of the waveform, with 100 kHz in the center of the display. The two sidebands are spaced by 10 kHz on each side of the carrier and their amplitude is about –12 dB relative to the carrier.

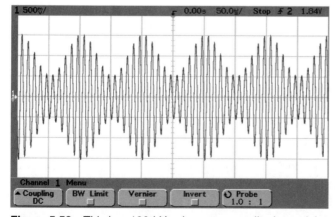

Figure 5.52 This is a 100-kHz sine wave amplitude modulated with a 10-kHz signal.

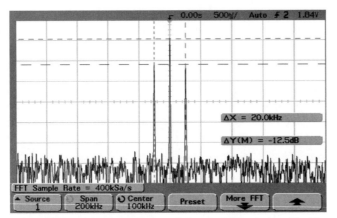

Figure 5.53 The FFT display of the modulated signal in Figure 5.52 shows the 100-kHz carrier and two sidebands 10 kHz away from the carrier.

5.14.5 FFT Summary

The two previous examples of FFT measurements illustrate the ability of the FFT function to determine the individual frequency content of complex waveforms. This is an important attribute of the FFT and can be used on many applications beyond harmonic distortion and amplitude modulation.

The FFT is a very powerful feature that lets the user gain new insight into the waveform that a time domain picture cannot deliver. However, because aliasing can occur with the FFT function, the user must be on guard when interpreting the results. Out-of-band signals that were not expected may show up as aliases and present confusing and misleading results.

5.15   Basic Time Domain Reflectometry

Time domain reflectometry (TDR) is a powerful measurement technique that uses the reflections of a pulse sent down a transmission line to characterize the impedance of that line. While specialty instruments have been created specifically for the purpose of performing TDR measurements, this section discusses how a general-purpose scope and pulse generator can be combined to deliver basic TDR measurements.

5.15.1 Transmission Line Theory

Transmission lines are often used to connect instruments to the circuit under test or to connect multiple instruments together. A common type of transmission line is the coaxial cable, often supplied with BNC connectors on each end for easy connection to test equipment. Transmission lines are constructed such that they present a consistent impedance to waveforms propagating down the transmission line. This impedance is called the *characteristic impedance*, Z_0, typically 50 Ω.

5.15.2 Propagation Velocity

Signals do not travel down a transmission line infinitely fast. It takes a finite amount of time for a signal to propagate from one place to another. For systems that have short cable length (relative to the wavelength of interest), these effects are usually ignored. As the frequency or speed of the signal and/or the length of the line is increased, the delays along the line become significant.

Electromagnetic waves in free space propagate at the speed of light. Inside a transmission line there is usually a dielectric material that lowers the propagation velocity to less than the speed of light. The *propagation velocity* is given by

$$v_p = k_v \cdot c$$

where k_v is the *velocity factor* and c is the velocity of light in free space (approximately 3×10^8 meters/second).

The velocity factor expresses the propagation velocity as a percent of the free space light velocity. The velocity factor has a value between 0 and 1 depending on the dielectric material in the transmission line. The cable manufacturer will usually specify the propagation velocity in the form of velocity factor, often expressed in percent. Typically, k_v ranges from 60% to 90%.

5.15.3 Source, Line, and Load

Let's first consider the source connected to a load, shown in Figure 5.54. The source voltage steps from zero volts to 1 volt. If the load was connected directly to the source with no transmission line in between, $v_1(t)$ and $v_2(t)$ would be identical. We would expect to see v_1 and v_2 instantly jump to 0.5 V, due to the voltage divider effect of the two Z_0 impedances. However, with the transmission line inserted, it will take a small amount of time for the voltage step to propagate down the line. When the source voltage changes from 0 volts to 1 volt, an incident voltage is created at the source end of the transmission line. Since the source sees the Z_0 impedance of the line, this incident voltage is equal to one-half of the source voltage. The incident voltage travels down the transmission line at the propagation velocity until it meets the load. Since the load impedance is equal to the characteristic impedance of the line, the situation is perfectly matched and no reflections occur. The incident voltage is "absorbed" by the load and $v_2(t)$ takes on the value of 0.5 V. There is a time delay between $v_1(t)$ and $v_2(t)$ due to the propagation time down the transmission line, as shown in Figure 5.54b.

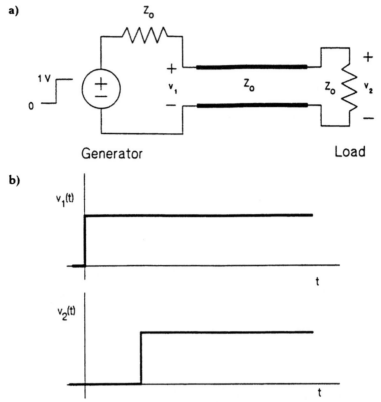

Figure 5.54 (a) A source with Z_0 impedance drives a Z_0 load via a Z_0 transmission line. (b) It takes a finite amount of time for the voltage to travel down the transmission line.

5.15.4 Reflection Coefficient

Now suppose the Z_0 load is replaced by a load that is some other value, such that it is not matched to the transmission line. As in the Z_0 load case, the incident voltage of 0.5 volts appears at the source end of the transmission line. The incident voltage is not affected by the change in load impedance since the source initially sees only the Z_0 impedance of the line. The 0.5-volt step propagates down the line and eventually reaches the load. The load is *not* matched to the Z_0 line, so some of the incident voltage is reflected back toward the source. The reflected voltage is given by

$$V_R = \rho \cdot V_I$$

where V_R is the reflected voltage, V_I is the incident voltage and ρ is the reflection coefficient.

$$\rho = \frac{V_R}{V_I}$$

The value of ρ is always between -1 and 1, inclusive. For the case shown, ρ can be computed from Z_0 and Z_L using the following equation.

$$\rho = \frac{Z_L - Z_0}{Z_L + Z_0}$$

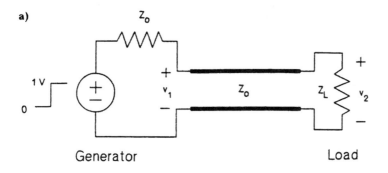

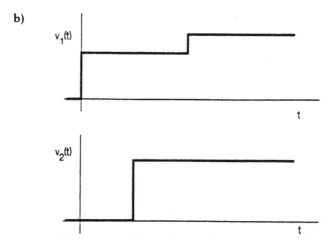

Figure 5.55 (a) A source with Z_0 impedance drives a non-Z_0 load via a Z_0 transmission line. (b) The incident wave appears at v_1, travels down the line to v_2, where a portion of the incident voltage is reflected back to the source. This reflected wave travels back to the source and appears at v_1.

The reflected voltage, V_R, propagates back up the line to the source. The voltage at any point on the line is the sum of the incident and reflected voltages, taking into account how far the two voltages have traveled at any point in time. The line is initially at 0 volts (because the source has presumably been at 0 volts for some time). As the incident wave travels down the line, the line becomes charged to V_I. Then the reflected wave starts back down the line moving from the load toward the source. As the wave passes any given point, the voltage on the line at that point goes from V_I to $V_I + V_R$. When the reflected wave reaches the source, it encounters the Z_0 impedance and no additional reflections occur. Had the source impedance been other than Z_0, additional reflections would occur.

Table 5.2   Table of Reflection Coefficient and Z_L/Z_0

Reflection Coefficient	Z_L/Z_0
−1.00	0.00
−0.60	0.25
−0.33	0.50
−0.14	0.75
0.00	1.00
0.20	1.50
0.33	2.00
0.67	5.00
0.82	10.00
0.90	20.00
0.96	50.00
1.00	∞

5.15.5 Matched Load

Our TDR test setup is shown in Figure 5.56. This test setup is equivalent to the diagram in Figure 5.54. The pulse generator is the source, the cable is the transmission line, and a Z_0 load is attached at the end of the cable. The scope is connected using a BNC tee such that it monitors the voltage at $v_1(t)$ in Figure 5.54. The pulse generator output imped-

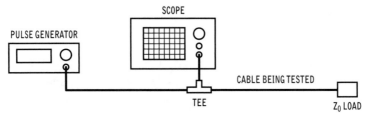

Figure 5.56 A pulse generator, a tee connection, and a scope are configured for TDR measurement of a cable.

ance, the cable, and the load are all nominal 50-Ω impedance. The scope uses its 1-MΩ impedance to monitor the voltage at the tee without loading down the system.

The set up of the pulse generator is not critical. The pulse amplitude is chosen to provide a large enough signal for easy viewing on the scope (in this case 1 volt). The pulse period must be longer than the time duration being measured.

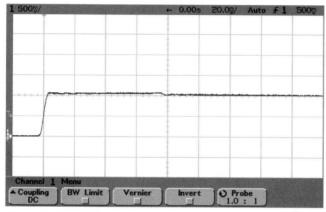

Figure 5.57 TDR measurement of a long cable terminated with 50 ohms.

The TDR measurement shown in Figure 5.57 shows that with the cable terminated in 50-Ω, the step voltage propagates down the line and does not reflect back. The cable being measured is 8 meters long and has a velocity factor of 66%. The 1V incident voltage travels at $0.66 \times 3 \times 10^8$ m/sec or 1.98×10^8 m/sec. Thus, it takes approximately 40 nsec for a voltage wave to propagate down the length of the line. Closer examination of Figure 5.57 reveals a small bump in the voltage 80 nsec (4 divisions) after the voltage step. This is a small reflection off the end of the cable, probably due to a slight mismatch between the load and the characteristic impedance of the cable. The effect of this mismatch shows up at *twice* the total delay time associated with the transmission line. This is because the step has to propagate down to the end of the cable and then propagate all of the way back before this small reflection is visible at the scope.

5.15.6 Open Circuit Load

Removing the load from the end of the cable gives us an open circuit condition at the load (Figure 5.58). The open circuit condition at the end of the cable produces a reflection coefficient of 1, so the entire incident voltage is reflected back to the source. This reflection *adds* to the incident voltage and creates a second step at 80 nsec after the initial pulse. While the small reflection of Figure 5.57 was rather subtle, here we clearly see the reflection at the end of the transmission line. The incident voltage is approximately 1 volt in amplitude and the reflected voltage bumps the voltage up to 2 volts.

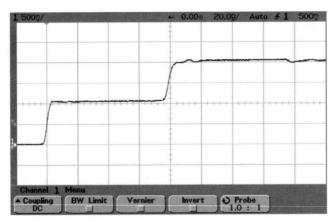

Figure 5.58   TDR measurement of a long cable terminated in an open circuit.

5.15.7 Short Circuit Load

Putting a short circuit on the end of the transmission line causes the reflection coefficient to be −1. The entire incident voltage is reflected back to the source, but with a negative polarity. At the 80-nsec mark, this causes the voltage to drop back to zero, with the incident and reflected voltages canceling.

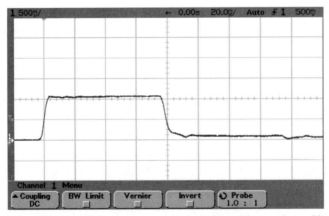

Figure 5.59   TDR measurement of a long cable terminated in a short circuit.

We can make a few general observations about the TDR measurements of Figures 5.57, 5.58, 5.59 (Z_0, open and short loads). The horizontal axis of the scope is measured in time but for a particular propagation velocity, time maps directly into distance down the transmission line. The vertical axis gives us a reading on the reflection coefficient or impedance on the line at a particular point. Together, these two axes give us a picture of the transmission line characteristics as a function of distance down the line. For example, in Figure

5.58, we see that the TDR plot shows no reflections for the first 80 nsec. During the 80 nsec, the waveform is at center screen at the voltage that indicates a Z_0 impedance (or $\rho = 0$). At the end of the line, we see the waveform rise to twice the $\rho = 0$ value, corresponding to an open circuit or $\rho = 1$. Similarly, in Figure 5.59, we see that the TDR plot starts out at the $\rho = 0$ (Z_0) level and then drops down to $\rho = -1$ (0 Ω).

5.15.8 Distance down the Transmission Line

Note that the point of a reflection on the line shows up consistently in the TDR measurement. In the example measurements, the 80-nsec time delay corresponds to whatever reflection occurs (or doesn't occur) at the end of the cable.

In general, whenever a reflection occurs on the transmission as shown on the TDR display, we can use the TDR display to determine where the reflection occurs physically on the line. The distance from the monitoring point at the scope to the point of mismatch is given by:

$$d = \frac{v_p T}{2}$$

where T is the transit time from voltage step to mismatch and back again (as measured on the scope).

5.15.9 Other Impedances

Figure 5.60 shows the TDR measurement with a 12.5-Ω load at the end of the transmission line. This load condition at the end of the cable produces a reflection coefficient of -0.6, so at the 80-nsec point, the incident voltage drops from 1 volt back down to about 0.4 volts. If we interpret the TDR display as a plot of ρ versus distance, we see that the transmission line starts out at $\rho = 0$. Then, 80 nsec later, the voltage drops to 0.4 volts, indicating a ρ of -0.6.

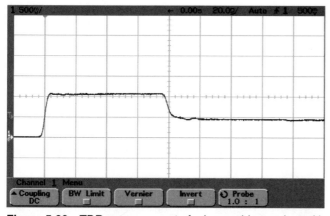

Figure 5.60   TDR measurement of a long cable terminated in 12.5 ohms.

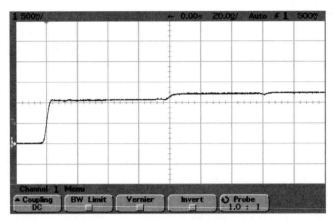

Figure 5.61   TDR measurement of a 50Ω, which is connected to a 75Ω cable that is terminated in 75Ω.

Figure 5.61 shows the TDR measurement of our same 8-meter 50-Ω transmission line connected to a 7.6-meter long 75-Ω transmission line, which is terminated in 75 Ω. The 75-Ω line has the same propagation velocity as the 50-Ω line (66%). The TDR measurement starts out at $\rho = 0$ for the first 80 nsec, then pumps up slightly where the 50 Ω connects to the 75-Ω line, indicating an increase in impedance. Since the 75-Ω line is terminated in its Z_0 load, there is no significant reflection at the end of the line.

Obviously, one application of the TDR measurement is to check the integrity of transmission lines. This measurement can identify short or open circuits on the line, or even variations in the impedance of the line. More importantly, we can measure the distance down the cable that a fault occurs, allowing for identification of the fault and repair of the cable.

The resolution and accuracy of the TDR measurement is limited by the combined rise time of the scope and pulse generator. It will be difficult to resolve any impedance changes or reflections any finer than this rise time. Therefore, higher bandwidth scopes and faster pulse generators can deliver finer resolution in terms of distance down the line.

5.16   References

"Agilent 54600-Series Oscilloscopes User's Guide," Agilent Technologies, Part No. 54622-97002, March 2000.

"Time Domain Reflectometry Theory," Application Note 1304-2, Agilent Technologies, Part No. 5966-4855, May 1998.

Brigham, E. Oran. *The Fast Fourier Transform and Its Applications*, Englewood Cliffs, NJ: Prentice-Hall, 1997.

Lenk, John D. *Handbook of Oscilloscopes*, Englewood Cliffs, NJ: Prentice Hall, 1982.

Oliver, Bernard M., and John M. Cage. *Electronic Measurements and Instrumentation*. New York: McGraw-Hill, 1971.

Ramirez, Robert W. *The FFT, Fundamentals and Concepts.* Englewood Cliffs, NJ: Prentice-Hall, 1985.

Wedlock, Bruce D., and James K. Roberge. *Electronic Components and Measurements.* Englewood Cliffs, NJ: Prentice Hall, 1969.

Witte, Robert A. *Spectrum and Network Measurements*, Norcoss, GA: Noble, 2001.

CHAPTER 6

Electronic Counters

This chapter discusses the electronic counter, commonly used for measuring frequency and time parameters of a signal. Electronic counters use stable crystal oscillators and digital counting circuits to provide simple and accurate frequency and period measurements. Some counters can measure additional parameters of a signal, including time internal, pulse width, rise time, fall time, and total number of events.

6.1 Basic Frequency Counter

A convenient and accurate way of measuring frequency is to use a *frequency counter*. The frequency of a periodic waveform is defined as the number of cycles that occur per second. A frequency counter uses a precise internal timebase and digital counters to produce a digital frequency readout.

A frequency counter measures the frequency of the signal over a defined time interval. To a frequency counter, the definition of frequency is

$$f = \frac{n}{T_C}$$

where n = the number of cycles of the waveform counted

T_C = the time interval over which the cycles are counted

Figure 6.1 shows a simplified block diagram of a basic frequency counter. The signal being measured is amplified and changed into a digital pulse train. This pulse train passes through an electronic switch called the *main gate* and drives a series of digital counters. If

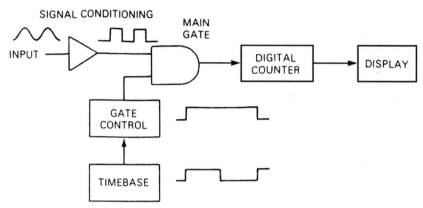

Figure 6.1   A simplified block diagram of a frequency counter. The number of cycles in the signal being measured is counted for a length of time determined by the timebase.

the main gate is open, the value of the digital counters increases by one for each new cycle of the signal being measured. If the main gate were to remain open, the digital counter would keep counting up indefinitely (or at least until it ran out of digits). Instead, the main gate is opened for a known length of time and the resulting number of cycles of the waveform is measured. This number represents the frequency of the waveform. To perform another measurement, the digital counter is reset and the main gate is once again reopened. The waveforms associated with this operation are shown in Figure 6.2.

As an example, suppose that the length of time that the main gate is opened is 1 second. Using this 1-second gate time means that the frequency in Hertz (which is the number of cycles in 1 second) will be displayed on the digital counter. This gate time is convenient for explaining the operation of the frequency counter. However, for high-frequency signals, the 1-second gate time would cause the digital counters to reach their maximum count and overflow. For these signals a shorter gate time is needed and for low-frequency signals a longer gate time is needed. These different gate times correspond to the different measurement ranges of the counter.

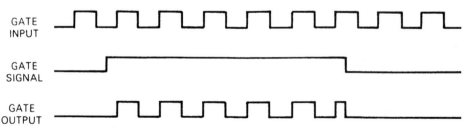

Figure 6.2   The signals associated with the block diagram of the frequency counter. The gate signal turns on the main gate, allowing the input signal to pass through to the digital counter.

In Figure 6.2, note that the last pulse at the end of the gate output is shorter than the rest of the pulses. This is due to the main gate closing midway through the period of the input signal. Had the gate closed slightly earlier, this pulse would have been completely suppressed. Had the gate closed later, the full pulse (and perhaps the start of the next input period) would have been passed on to the digital counter. This illustrates the ±1 count ambiguity that is typical of counter measurements. That is, a very slight variation in the waveform can cause the count to change by one count.

A typical frequency counter is shown in Figure 6.3.

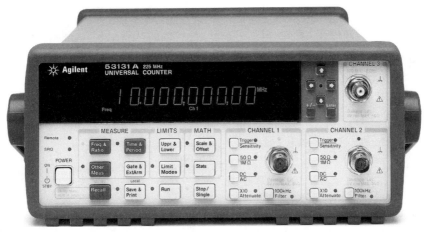

Figure 6.3   A 225-MHz universal counter. (Photo: Agilent Technologies. Reprinted with permission.)

Example 6.1

A basic frequency counter counts 400,000 cycles of a waveform over a time interval of 250 msec. What is the frequency of the waveform?

$$f = \frac{n}{T_C} = \frac{400000}{0.25} = 1.6 \text{ MHz}$$

6.2   Frequency Dividers

A *frequency divider* is used to reduce the frequency of a digital signal. Figure 6.4a shows a frequency divider, which divides the input frequency by 2 to produce an output frequency at half the input frequency. Similarly, Figure 6.4b shows a divide-by-10 circuit, which reduces its input frequency by a factor of 10. Notice that both the input and output of the divider is a pulse train. This type of circuit can be used to increase the range of a frequency counter in two ways.

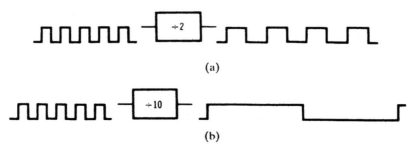

(a)

(b)

Figure 6.4   A frequency divider produces an output frequency that is equal to the input frequency divided by an integer number. a) A divide-by-2 frequency divider. b) A divide-by-10 frequency divider.

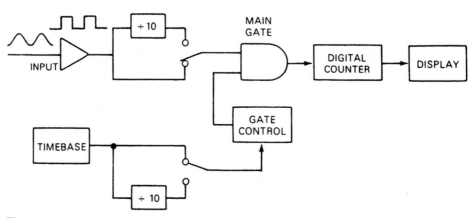

Figure 6.5   A block diagram of a frequency counter showing the use of frequency dividers to increase the frequency range of the basic counter.

Figure 6.5 shows a divide-by-10 circuit added to the frequency counter in two different places. At the input, the frequency divider has the effect of increasing the maximum measurable frequency by reducing incoming signals to a range that is usable by the basic frequency counter. Used in this manner, frequency dividers are often referred to as *prescalers*. For instance, the basic frequency counter may have a maximum frequency limitation of 100 MHz. Adding a divide-by-10 circuit in front of the basic counter extends the range by a factor of 10, to 1,000 MHz. In this case, the prescaler also causes the resolution of the measurement to be reduced by a factor of 10.

Another frequency divider is used to divide the frequency of the timebase circuit. This has the effect of increasing the length of time that the main gate is on, which means that lower frequencies can be measured than with the original frequency counter. Another benefit of the increased gate time is improved frequency resolution.

In both cases, the frequency divider is shown as being able to be switched in and out of the circuit. Typically, several divider circuits are supplied so that the user can conveniently select a measurement range on the instrument. Sometimes a special high-frequency prescaler is offered as an external option that increases the high-frequency range of a counter.

6.3   Period Measurement

The user of a frequency counter can calculate the period of a waveform based on its frequency ($f = 1/T$), but period measurement is often built into a frequency counter. Figure 6.6 shows a small but important change in the basic frequency counter block diagram—the input and timebase connections are interchanged. In this mode, the input opens the main gate for one of its cycles. During this cycle (which is the period of the input), the number of timebase clocks are counted. Suppose the timebase period was 1 msec, then the resulting display would be the number of 1-msec cycles that occurred during one cycle of the input waveform. In other words, this represents the period of the input waveform in msec. By using selectable frequency dividers, other timebase frequencies can be generated, resulting in other ranges of period measurement.

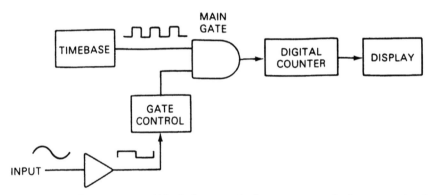

Figure 6.6   The conceptual block diagram of a frequency counter for measuring the period of the input waveform.

6.4   Reciprocal Counter

The two block diagrams associated with frequency measurement and period measurement each have their own advantages and disadvantages. For low frequencies, period measurement has higher resolution and is faster. For frequencies above the timebase clock, the frequency measuring block diagram is better. Considering numerical examples will help highlight the advantages and disadvantages of each counting technique.

6.4.1   Low Frequency

Suppose a counter has a timebase clock of 10 MHz and can operate in either the frequency or period mode. First, consider the case where the input frequency is very low, for example, 1 Hz. In the period mode, the counter can produce the result in one cycle of the input waveform, in this case, 1 second. The period measurement has an uncertainty of ±1 count of the 10-MHz timebase, or ±100 nsec. In frequency mode, the counter counts the number of input cycles that occur within the measurement time. As an example, let's assume that the measurement time is 10 sec (which happens to be much longer than the time the period measurement took). During 10 seconds, approximately 10 input cycles occur and the counter measures this frequency with an uncertainty of ±1 count, or ±1 Hz. The result is that even though the frequency mode takes much longer to complete a measurement, the frequency measurement is much less accurate than the period mode.

6.4.2   High Frequency

Now consider the case where the input frequency is much higher than the timebase frequency. For example, say the input frequency is 100 MHz, 10 times higher than the timebase frequency. If measured with the period mode of the counter, the 10-nsec period is measured with an uncertainty of ±1 timebase count, or ±100 nsec. In frequency mode, the uncertainty is again ±1 count of the input frequency, or ±10 nsec. For frequencies higher than the timebase frequency, the frequency mode is more accurate.

6.4.3   Reciprocal Counter

The *reciprocal counter* combines the frequency and period modes and is implemented as shown in Figure 6.7. The block diagram contains two counters: the *time counter* and the *event counter*. The time counter counts the timebase periods while the event counter counts the cycles of the input frequency. The period of the input signal is given by

$$T = \frac{T_{CLK} N_{CLK}}{N_{EVENT}}$$

where T_{CLK} = the timebase clock period

N_{CLK} = the clock count

N_{EVENT} = the event count

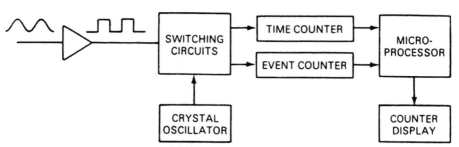

Figure 6.7   The reciprocal counter combines the best features of frequency mode and period mode.

Suppose the reciprocal counter is configured such that the gate time is one period of the input signal. The event counter records a count of 1, while the time counter counts the number of reference clock periods that occurred during the gate time. The period measurement is obtained by multiplying the number period of the reference clock times the time counter, divided by the event count of 1. If the reciprocal counter is configured to measure over many cycles of the input signal, the event counter counts these cycles and is used to calculate the average period of the waveform during the gate time.

The reciprocal counter requires some computational power in the instrument, usually in the form of a microprocessor. Most modern test instruments include at least one microprocessor for control of the instrument anyway, so this is not a problem.

Example 6.2

A reciprocal counter that has a timebase clock frequency of 100 MHz counts 50 timebase periods during 20 cycles of the input signal. What is the frequency of the signal?

T_{CLK} = 1/100 MHz = 10 nsec

N_{CLK} = 50

N_{EVENT} = 20

$$T = \frac{T_{CLK} N_{CLK}}{N_{EVENT}} = (10\text{nsec } 50)/20 = 25 \text{ nsec}$$

$$f = \frac{1}{T} = 40 \text{ MHz}$$

Basic electronic counters employ digital counters that count up the number of cycles of the timebase and the input signal. The size of the counter limits the resolution of the measurement. Since the end of an input cycle will typically not occur right at the end of a timebase period, one clock cycle may just barely be missed. This determines the resolution of the measurement. One way to improve the resolution is to use *interpolation* circuitry that measures the time from the event of interest to the start of a timebase period. This interpolated value is included in the calculation of the counter measurement.

6.5 Universal Counter

Universal counters extend the basic counter circuitry such that it can measure other waveform parameters. Besides frequency and period, measurements commonly made by universal counters include time interval, pulse width, duty cycle, rise time, fall time, and phase. Time interval measurement and pulse width measurement are a logical extension of a period measurement, with the start and stop times of the measurement no longer limited to the start and stop of one period. Rise time is derived from the time interval circuitry, but with the start and stop times determined by comparators, which trigger at the 10% and 90% points of the waveform (Figure 6.8). Duty cycle is a pulse width measurement divided by the period of the waveform. A phase measurement measures the time delay between two waveforms and expresses in degrees, with 360 degrees corresponding to one period of the waveform. Table 6.1 summarizes the measurements typically found in universal counters.

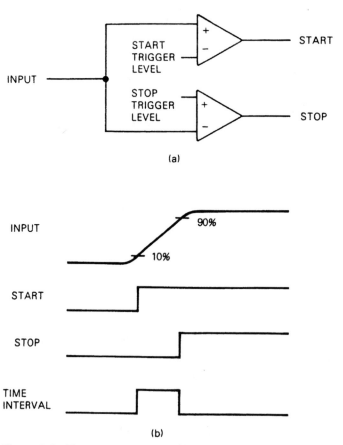

Figure 6.8 Two comparators are used to convert the rise time of the input signal to a time interval that is measurable by counter circuits. a) The comparator circuit. b) The signals associated with the rise time measurement.

Table 6.1   Universal Counter Measurements

Measurement	Description
Frequency	Single-channel frequency measurement.
Frequency ratio	The ratio of the frequencies of two channels.
Time interval	The time difference between two edges, on the same waveform or different waveforms.
Period	Single-channel period measurement.
Rise/fall time	The rise or fall time of a waveform (usually the 10% and 90% level).
Positive/negative pulse width	The width of a positive or negative pulse, usually measured at the 50% point on the waveform.
Duty cycle	The positive pulse width of a waveform divided by the period of the waveform.
Phase	The time difference between two different waveforms, expressed in degrees with the period of one of the waveforms defined as 360 degrees.
Totalize	The total number of events counted on a waveform.
Peak voltage	The peak positive or negative voltage of the waveform.

6.6   Gated Counter Measurements

For signals such as pulsed RF waveforms, the sinusoid being measured pulses on and off. Attempts to measure the frequency of such a signal by measuring the number of cycles for a time longer than the pulse width will be in error. A gated counter measurement provides control of the portion of the waveform where the frequency count is taken. Counters with gated measurements automatically gate the counter circuitry on for a controlled length of time and perform the measurement only while the pulsed RF is active (Figure 6.9).

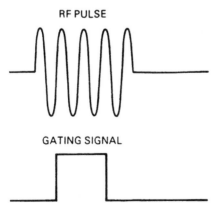

Figure 6.9   A frequency counter with a gated measurement mode can measure the frequency during an RF pulse.

6.7   Timebase Accuracy

The timebase of a frequency counter is usually a precisely controlled crystal oscillator, which is divided down to produce the required reference frequency. This results in a timebase accuracy that is limited only by the stability and accuracy of the crystal oscillator. Since this oscillator must operate at only one frequency, it can be designed to be extremely stable.

There are three main types of crystal oscillators: *room temperature crystal oscillator (RTXO), temperature-compensated crystal oscillator (TCXO), and oven-controlled crystal oscillator.* Room temperature crystal oscillators are designed to be relatively stable over some temperature range, typically 0 to 50°C. A careful choice of crystal typically results in a temperature stability of 2.5 parts per million (ppm) over this temperature range. A TCXO takes the basic crystal oscillator design and adds a temperature-sensitive component or components, which compensate for the inherent temperature characteristics of the crystal. This can produce an order of magnitude improvement in stability, typically 0.5 ppm over the same temperature range. The ultimate solution to frequency stability is to remove the temperature variation that the crystal experiences. An oven-controlled crystal oscillator has the crystal housed in an oven, which is stabilized in temperature by a heating element. The control system for the heater can be a simple ON/OFF type or a more complex linear control system. Oven oscillators can typically achieve frequency stability of 10 parts in 10^9 or 0.01 ppm over a 0 to 50°C temperature range. Table 6.2 gives the typical frequency stability for the three main types of crystal oscillators.

Table 6.2   Typical Frequency Stability for Oscillators

Oscillator Type	Frequency Stability
RTXO	2.5 ppm
TCXO	0.5 ppm
Oven oscillator	0.01 ppm

Many counters have an input that allows an external frequency reference to be used. The counter locks its internal oscillator to the supplied reference (usually 10 MHz). In this mode, the stability of the counter is determined by the external reference.

6.8   Input Impedance

Frequency counters generally have either a 50-Ω or a 1-MΩ impedance. Like several of the instruments already discussed, the 1-MΩ input is convenient for general-purpose, low-frequency measurements. The 50-Ω becomes necessary as higher frequency measurements are made (typically greater than 50 MHz). For frequency and period measurements, the counter is interested only in the timing characteristics of the signal (and not amplitude or voltage). A decreased signal level does not necessarily impact measurement accuracy.

Thus, capacitive loading due to the 1-MΩ input may reduce the signal level into the frequency counter somewhat without introducing error into the measurement. At some point, however, the signal is attenuated so much that the counter can no longer detect the zero-crossings of the signal, making the frequency measurement inaccurate. Various combinations of 50-Ω and 1-MΩ inputs exist. Some counters include both, with the 1-MΩ input intended for low-frequency operation and the 50-Ω input for high-frequency operation. Other counters have one input that is switchable between 1-MΩ and 50-Ω.

6.9 Frequency Counter Specifications

The specifications for a typical frequency counter are given in Table 6.3. This counter has two inputs (A and B) and the frequency range is shown as dependent on which input is being used. The sensitivity specification defines how small of a signal can be reliably measured. The frequency resolution specification determines the smallest change in frequency that can be detected.

Table 6.3   Abbreviated Specifications for a Frequency Counter

Specification	Input A	Input B
Frequency range	10 Hz to 100 MHz	90 MHz to 1,000 MHz
Input impedance	1 MΩ, 25 pF	50 Ω
Sensitivity	15 mV RMS	10 mV RMS
Frequency resolution	8 digits	
Temperature stability (Standard timebase)	< 2 ppm, 0 to 40°C	
Temperature stability (Optional oven timebase)	< 0.1 ppm, 0 to 50°C	
Aging rate (Standard timebase)	< 0.1 ppm/month	
Aging rate (Optional oven timebase)	< 0.03 ppm/month	

The most desirable specification is conspicuous by its absence. There is no frequency accuracy spec. This seems a bit strange at first, until one understands what limits the performance of the instrument. If the timebase of a frequency counter is adjusted to be exactly on frequency, presumably by comparing it to some perfect frequency standard, then there is essentially no frequency error at that instant in time. However, over any time interval the timebase tends to drift in frequency, primarily due to aging effects in the crystal oscillator and changing performance with temperature. Fortunately, the aging and temperature stability are usually specified by the manufacturer.

Consider the typical specifications given in Table 6.3. Ignoring temperature changes for the moment, the long-term frequency stability (with the standard time base) is determined by the aging rate (0.1 ppm/month). Assuming the frequency counter was adjusted to be exactly on frequency (to within the resolution of the counter) at some point in time, then one month later it may be off by as much as 0.1 parts per million. For a frequency of 100 MHz, the maximum error would be 10 Hz. One year later, the frequency could be off by as much as 12×0.1 ppm = 1.2 ppm. This results in an error of 120 Hz on a 100-MHz signal. Similar calculations can be made for the effect of temperature stability.

Example 6.3

A universal counter that is measuring a 30-MHz signal has the specifications shown in Table 6.3 (with standard timebase). How far can the frequency measurement be off due to temperature if the ambient temperature is 0°C?

The temperature stability is 2 ppm for the temperature range of 0 to 40°C.

The 300 MHz signal can be in error due to temperature by

$f_{error} = (2 \times 10^{-6})\ 300\ \text{MHz} = 600\ \text{Hz}.$

6.10   Time Interval Analyzer

Most frequency measurements are made on stationary sine waves such that the frequency is not varying over time. In the case of pulsed RF, the sine wave is present only part of the time so the counter must measure the frequency during the pulse ON time. This is a simple example of a signal that is not one stable frequency, but changes over time. Other examples are even more complex. For instance, a voltage-controlled oscillator (VCO) circuit might be designed to sweep in frequency, starting at f_1 and sweeping linearly to f_2 (Figure 6.10). With such a measurement, a normal frequency counter would be confused and would give a reading that represents the average frequency over some segment of time.

To handle complex signals such as the sweeping VCO, the *time interval analyzer* (also known as the *modulation domain analyzer* or *frequency and time interval analyzer*) was invented. This class of instrument measures the instantaneous frequency as a function of time. This may seem confusing at first, since we are more used to waveforms being represented by voltage as a function of time. Frequency domain measurements made with spectrum analyzers display signals as voltage versus frequency. But frequency versus time? This is something new. Returning to the case of the sweeping VCO, it is easy to understand the need for such an instrument. How could an instrument user tell how linear the VCO sweep is? While it is possible to use an oscilloscope display to capture the voltage versis time information and convert it to frequency versus time, the memory size of the scope must be very large and the timing resolution must be very fine. A more direct and efficient way to make this measurement is to use an instrument that measures the time between zero-crossings on the waveform and plots these as a function of time. Thus, a time interval analyzer can plot the waveform's period versus time, frequency versus time, phase versus time, or time interval versus time.

The time interval analyzer is essentially a frequency counter that can produce a real-time frequency reading, which is plotted as a function of time. The key component in the modulation domain analyzer is the *zero dead time (ZDT) counter*, which can track the input signal's frequency and output the resulting value without missing a period of the signal. Otherwise, the counter would miss portions of the waveform, giving a discontinuous frequency plot.

The sweeping VCO represents a common use of the time interval analyzer. Another application is the measurement of time jitter and frequency modulation on an otherwise stable carrier. Some spread spectrum radio transceivers use frequency hopping techniques to reduce the susceptibility to jamming and other interference. Time interval analyzers are very effective at tracking the frequency of the receiver's local oscillator as it jumps from place to place.

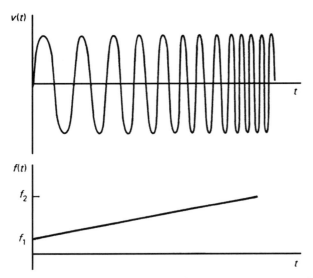

Figure 6.10 The frequency of an oscillator that sweeps from f_1 to f_2 can be tracked as a function of time by a modulation domain analyzer.

6.11   References

"53131A/132A 225 MHz Universal Counter Operating Guide," Agilent Technologies, May 1999.

Coombs, Clyde F., Jr. *Electronic Instrument Handbook*, 3rd ed., New York: McGraw-Hill, 1999.

"Fundamentals of Electronic Counters," Hewlett-Packard Co., Application Note 200, Publication No. 5952-7506, July 1978.

"Fundamentals of Microwave Frequency Counters," Agilent Technologies, Application Note 200-1, Publication No. 5965-7661E, May 1997.

Helfrick, Albert D., and William D. Cooper. *Modern Electronic Instrumentation and Measurement Technique*, Englewood Cliffs, NJ: Prentice Hall, 1990.

Oliver, Bernard M., and John M. Cage. *Electronic Measurements and Instrumentation*, New York: McGraw-Hill, 1971.

CHAPTER 7

Power Supplies

Electrical power is required by all electronic circuits to function properly. While electricity is commonly distributed as an AC voltage, most electronic circuits require a DC voltage to operate. This can be supplied by a battery, a built-in power supply or a bench power supply. For portable applications, a battery may be a good solution. For circuits that are used where AC power is available, having a built-in power supply makes sense. But when designing or testing a circuit, a bench DC supply will usually be the most convenient.

The use of a bench power supply may simply be a way to supply the right voltage to a circuit. In this case, the user sets the power supply to the right voltage and leaves it there. Alternatively, the power supply voltage may be varied as part of the test procedure to evaluate how a device responds to variation in supply voltage or current.

7.1 Power Supplies

A *fixed* power supply is designed to operate at a particular constant output voltage. An adjustment may be included on a fixed power supply to precisely set the output voltage. Since the adjustment range is usually only 5 or 10% around the nominal value, the supply is still considered to have a fixed voltage.

Variable power supplies can be adjusted to produce a wider range of voltages, typically 0 to 20 volts or more. This type of supply is more versatile since the voltage can be set to match the particular circuit requirement. In addition, the voltage can be varied around the nominal value during design and testing.

7.2 Circuit Model

The simplest circuit model for the DC power supply is just a DC voltage source (Figure 7.1a). With this model, no matter what else happens, the voltage across the two terminals is always the DC value of the voltage source. This is precisely what is desired in a voltage source—a constant DC voltage—and this circuit model is valid for many applications.

A more realistic circuit model, including the internal resistance of the power supply, is shown in Figure 7.1b. This resistance corresponds to the phenomenon that as current is drawn from the power supply, the output voltage decreases. For a high-quality power supply, this resistance value is very small (typically less than an ohm).

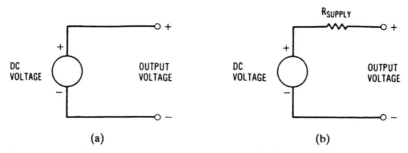

(a)   (b)

Figure 7.1 Circuit models for a DC power supply. a) The simplest model. b) A model including the internal resistance of the power supply.

Just as with other tests instruments, the grounding of a power supply output must be understood. Figure 7.2 shows the output connectors of a typical power supply. Three terminals are provided: the + and − terminals are the two connections for the output voltage and the third terminal is the ground connection. If the ground connection is left disconnected, the power supply's output is floating. That is, both the + and the − terminals can be connected anywhere in the circuit without regard to grounding. If the ground terminal is connected to the + or − terminal, then the power supply is grounded. In this way, a floating power supply can be configured to be positive or negative with respect to ground. A grounding strap that connects the two terminals is sometimes supplied to make this convenient. Some supplies are inherently grounded and will not allow the negative terminal to be at any electrical potential other than ground.

Many analog electronic circuits such as operation amplifiers require both positive and negative power supply voltages. These two supplies are often built into one *dual-output supply.* The circuit in Figure 7.3 shows how the two supplies are connected internally. The supplies shown are floating but can be grounded as needed by connecting the common terminal

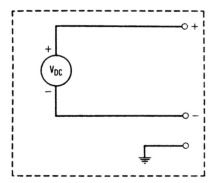

Figure 7.2 Circuit model for a floating power supply with optional ground connection.

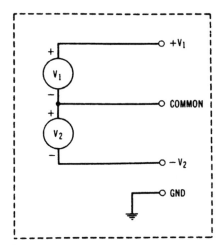

Figure 7.3 Circuit model for a dual floating power supply.

to the ground terminal. Figure 7.4 shows how two single power supplies can be connected to produce a dual power supply. At least one of the power supplies must be floating to avoid any conflicts in grounding.

A common configuration for a power supply is to have three outputs, with two of the outputs intended to power analog circuitry and the third output to support digital circuits. This *triple-output supply* is the most common configuration for a general-purpose power supply used for mixed analog and digital circuitry (Figure 7.5). Typically, the two analog-oriented outputs can produce up to 25 volts, while the digital-oriented supply tops out at 6 volts. The digital output can usually source significantly more current than the other two outputs, consistent with the higher current demand of digital circuits.

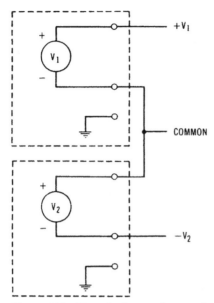

Figure 7.4   Two single supplies can be connected to create a dual supply.

Figure 7.5   A triple output programmable power supply with digital display. (Photo: Agilent Technologies. Reprinted with permission.)

The circuit diagram for a simple power supply is shown in Figure 7.6. The AC line voltage is connected to the transformer, which steps down the voltage to something closer to the final DC value. The AC line voltage and frequency varies around the world (some representative countries are listed in Table 7.1). The diode (or rectifier) changes the AC sine wave voltage into a half sine wave. The filter capacitor smooths out the half sine to approximate a DC

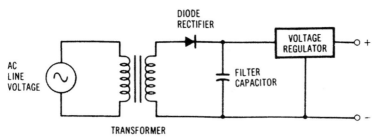

Figure 7.6   Circuit diagram for a simple DC power supply.

voltage. This DC voltage is not well controlled so it is passed through a circuit called a *voltage regulator,* which precisely controls the output voltage. This is one of the simplest types of power supplies, shown here to briefly introduce the steps necessary to change AC into DC.

Table 7.1   Standard AC Power in Selected Countries

Country	Nominal Voltage (RMS)	Frequency
Canada	120 volts	60 Hz
China	220 volts	50 Hz
France	220 volts	50 Hz
Germany	230 volts	50 Hz
Japan	100 volts	50 Hz and 60 Hz
Korea	220 volts	60 Hz
Singapore	230 volts	50 Hz
Taiwan	110 volts	60 Hz
United Kingdom	230 volts	50 Hz
United States	120 volts	60 Hz

Source: Adapted from *Electric Current Abroad*, U.S. Department of Commerce, 1997.

7.3   Constant-Voltage Operation

Most of the time, power supplies are used to produce a constant output voltage. The current supplied depends on the load applied to the power supply. This type of power supply use is referred to as *constant-voltage operation.* Figure 7.7 shows the voltage versus current plot of an ideal power supply in constant-voltage operation. The plot is a simple horizontal line, as the voltage is constant independent of the current supplied to the load.

If the load is a resistor, R_L, the output current conforms to Ohm's Law.

$$I_{OUT} = \frac{V_{OUT}}{R_L}$$

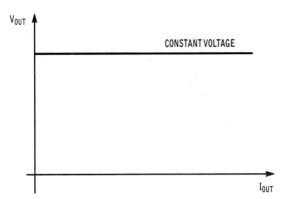

Figure 7.7 The voltage and current characteristics of an ideal power supply operating in constant-voltage mode.

7.4 Constant-Current Operation

Constant-current operation is the mirror image of constant-voltage operation. Here, the current supplied remains constant independent of the voltage at the power supply output. Figure 7.8 shows the voltage versus current plot of an ideal power supply in constant-current operation. The plot is a simple vertical line, as the current is constant independent of the voltage supplied to the load.

If the power supply is loaded by a resistor, R_L, the resulting voltage will obey Ohm's Law:

$$V_{OUT} = I_{OUT} \cdot R_L$$

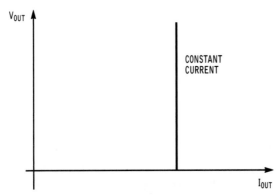

Figure 7.8 The voltage and current characteristics of an ideal power supply operating in constant-current mode.

7.5 CV-CC Operation

In normal operation, power supplies exhibit behavior that is a combination of constant voltage (CV) and constant current (CC) operation. In a typical power supply application, the power supply is set to a particular voltage, V_S, with a current limit of I_S. Figure 7.9 shows the plot of allowable operating range of the power supply. If the power supply is lightly loaded and is supplying a low current, the supply operates at location 1 in the figure. The supply is able to provide the required voltage and is operating well below the current limit.

As the load is increased on the supply and I_{OUT} increases, the operating point moves to the right, toward location 2. When the current reaches I_S, the power supply is operating with $V_{OUT} = V_S$ and $I_{OUT} = I_S$, indicated by location 2 in the figure. Any increase in current requirement from the load will cause the current limit to be activated. That is, the power supply will refuse to supply current larger than I_S, and consequently will reduce V_{OUT}. The operating point moves down toward location 3 in the figure, corresponding to a decrease in voltage, while the current remains equal to I_S.

The two lines in Figure 7.9 are called the *CV operating line* and the *CC operating line*. While the operating point remains on the CV line, the power supply is maintaining a constant voltage and responds to changes in load with a change in current. When operating on the CC line, the current remains constant and changes in load require a change in voltage.

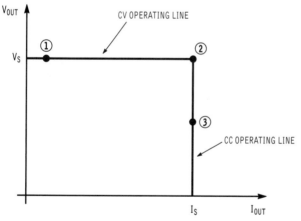

Figure 7.9 The voltage and current characteristics of a real power supply.

In most applications, we think about the power supply voltage setting, V_S, as the *desired* voltage output, while I_S is thought of as the current limit. A specific voltage value is required but any value of current is acceptable as long as it does not exceed the current limit. Figure 7.9 shows us that the voltage and current situation is symmetrical, with V_S and I_S being the voltage limit and current limit settings, respectively. V_S defines the maximum allowable voltage from the supply, just as I_S determines the maximum allowable current.

Example 7.1

A power supply is configured for a output voltage of 15 V with a current limit of 2A. What is the smallest value resistor (maximum load) that can be applied while still keeping the power supply in constant voltage operation?

The upper limit of constant-voltage operation occurs just as the current reaches the current limit. The resistor that produces this condition is

$$R_L = \frac{V_S}{I_S} = \frac{15V}{2A} = 7.5\Omega$$

7.6 Overvoltage/Overcurrent Protection

A good power supply provides built-in current limiting and short-circuit protection. This means that if the output of the supply is accidentally shorted out (and it will be), the supply reduces its voltage or shuts down the output completely to prevent excessive amounts of current to be drawn. This protects the supply itself and, to a lesser extent, the circuit under test.

Power supplies used on the test bench or in test systems may provide programmable *overcurrent protection* (OCP) and *overvoltage protection* (OVP). These are protection features that go beyond the voltage limit and current limit previously described. The OCP and OVP levels are normally set based on the power requirements and the damage level of the device under test. These limits should be set high enough to allow proper circuit operation (with some margin). That way, normal circuit operation is allowed but circuit faults will not draw any more current than necessary. When the output voltage or current exceeds the OVP or OCP level, the power supply output is shut down. Since tripping the OVP or OCP is considered a fault condition, the power supply will usually remain latched off until it is manually reset by the instrument user. Note that this is fundamentally different than the limits associated with constant-voltage and constant-current operation, where the supply is intentionally operated at the limit.

The user of a power supply should check the user's manual of the supply to understand the specific details of how the supply operates with regard to overvoltage and overcurrent protection.

7.7 Remote Sensing

Power supplies generally control their output voltage right at the output terminals of the supply. When external wires are used to connect to the device being powered, a voltage drop can occur across those wires. Figure 7.10a shows a load, R_L, connected to a power supply using wires that each have a resistance of R_{WIRE}. The voltage that is delivered to load is:

$$V_L = V_S - 2 \cdot I_{OUT} \cdot R_{WIRE}$$

For applications where the current is high or the length of wire must be long, the voltage drop can be significant. To remedy this situation, many power supplies provide *remote sensing* inputs that are used as shown in Figure 7.10b. The principle of operation is that the sensing inputs draw little or no current so that V_L is present at the sensing inputs without any voltage drop due to resistance in the wires. A voltage drop will still exist across R_{WIRE} but the power supply voltage, V_S, is automatically adjusted so that the desired voltage is maintained at V_L.

Remote sensing can correct for only a specified amount of loss in the wires, typically 0.5 to 1V per wire.

a)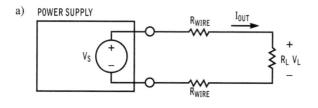

b)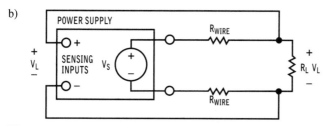

Figure 7.10 a) The resistance of the wires used to connect to the device being powered can cause a significant voltage drop. b) Remote sensing of the voltage is used to eliminate this voltage drop.

Example 7.2

A power supply uses remote sensing to ensure that 5.0 volts is delivered to the device being powered. The resistance of the wires connecting the power supply to the device is 0.1 Ω and the current supplied is 4A. Find the voltage delivered to the load (V_L) and the voltage at the output terminals of the supply (V_S).

Assuming that the remote sensing is working properly, V_L will be equal to the desired voltage, 5 volts. To achieve this, V_S will be higher than V_L by just enough voltage to offset the voltage drop across the two wires that connect the supply to the device.

$$V_S = V_L + 2 \cdot I_{OUT} \cdot R_{WIRE} = 5.0 + 2 \cdot 4 \cdot 0.1 = 5.8 \text{ volts}$$

7.8   Measurement Capability

While the primary function of a power supply is to provide a controlled voltage or current to a device under test, some power supplies include the ability to measure the voltage and current delivered by the supply. These measured values may be displayed on the front panel of the supply via a digital or analog meter. Monitoring the voltage and current during normal use of the power supply is very helpful and keeps the user informed of any unusual operating conditions. For example, if a power supply lead accidentally becomes disconnected, a quick check of the current meter will show that the output current has dropped to zero.

Programmable power supplies may have the ability to transfer voltage and current readings to a controlling computer, sometimes called *readback capability.* In a test system, this is a very convenient method for monitoring the voltage or current supplied to the device under test. Without this feature, an additional instrument such as a multimeter would need to be added to the test system to monitor the voltage or current. We can imagine a very simple test system consisting only of a computer and a programmable power supply that can measure the current versus voltage characteristics of a device under test.

7.9   Power Supply Specifications

There will be imperfections in the DC voltage produced by a power supply. The DC voltage may vary as the amount of current drawn from it changes and there may be a small amount of *AC ripple* remaining riding on top of the DC (Figure 7.11). The AC ripple is a remnant of the AC line voltage, which is not removed by the filter capacitor and voltage regulator. The ripple will be at the line frequency (60 Hz in the United States) plus the harmonics of the line frequency (120 Hz, 180 Hz, etc.).

A switching power supply regulates the output voltage by chopping or switching an unregulated DC voltage. This is an efficient type of regulation, but it can introduce additional frequencies in the DC output (typically 50 kHz to 1 MHz). The total ripple and noise

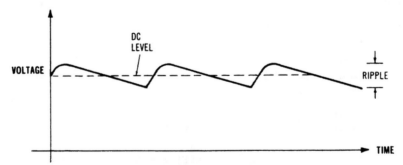

Figure 7.11   The DC output voltage of a power supply will have some small amount of AC ripple present.

in a power supply output is specified in terms of RMS voltage, peak-to-peak voltage, or both. In a quality power supply, these noise and ripple components will be low enough that they can usually be neglected for most general-purpose use.

The regulation specifications of the power supply describe how much the output voltage may vary under operating conditions. *Load regulation* (also called *load effect*) is how much the supply voltage varies with changes in power supply load.

$$\text{LOAD REG} = V_{\text{NO LOAD}} - V_{\text{FULL LOAD}}$$

$V_{NO\ LOAD}$ is the output voltage without a load connected and $V_{FULL\ LOAD}$ is the output voltage at the maximum load current (full load). *Line regulation* (also called *source effect*) is the amount that the output voltage varies due to changes in the power line voltage (over a specified range). Both types of regulation may be expressed as a voltage or as a percent of the output voltage.

Table 7.2 shows the specifications of a typical single-output power supply. Note that the output operates in two ranges; one delivering higher voltage but less current, the other delivering higher current with lower voltage. The basic power output capacity of the supply is about 120 W and the two operating ranges allow that power to be used most effectively for a particular application.

Table 7.2 Abbreviated Specifications of a Single-Output Programmable Bench Power Supply

Specification	Type	Value
Output ratings	Low range	0 to 15 V, 0 to 7A
	High range	0 to 30 V, 0 to 4A
Load regulation	Voltage	<0.01% + 2 mV
	Current	<0.01% + 250 μA
Line regulation	Voltage	<0.01% + 2 mV
	Current	<0.01% + 250 μA
Ripple and noise (20 Hz to 20 MHz)	Voltage	<0.35 mV RMS < 2 mV p–p
Programming accuracy	Voltage	± (0.05% + 10 mV)
	Current	± (0.2% + 10 mA)
Readback accuracy	Voltage	± (0.05% + 5 mV)
	Current	± (0.15% + 5 mA)

7.10   References

Coombs, Clyde F., Jr. *Electronic Instrument Handbook*, 3rd ed., New York: McGraw-Hill, 1999.

"E3632A DC Power Supply User's Guide," Agilent Technologies, Publication No. E3632-90002, April 2000.

"Electric Current Abroad," U.S. Department of Commerce, International Trade Administration, 1997, http://www.ita.doc.gov/td/machinery/ecabroad/

"Power Products Catalog 2001," Agilent Technologies, Publication No. 5968-2199E, February 2001.

Wolf, Stanley. *Guide to Electronic Measurements and Laboratory Practice*, Englewood Cliffs, NJ: Prentice-Hall, 1973.

CHAPTER 8

Spectrum and Network Analyzers

In the preceding chapters, waveforms have been usually character-ized and described in the time domain. As demonstrated in Chapter 1, these same waveforms can be analyzed and measured in the fre-quency domain. This frequency domain representation is also referred to as the *frequency spectrum* of the signal. Of primary interest is the *spectrum analyzer*, also referred to as a *signal analyzer*, which charac-terizes a waveform in the frequency domain the same way an oscillo-scope characterizes a waveform in the time domain. Closely related to the spectrum analyzer is the *network analyzer*, which is used to char-acterize the frequency response of electronic networks. *Distortion analyzers* are used for measuring distortion in audio circuitry. *RF power meters* are used to accurately measure the power level of radio frequency and microwave signals.

8.1   Spectrum Analyzers

Recall from Chapter 1 that a signal can be viewed in the time domain (Figure 8.1a) or the frequency domain (Figure 8.1b). The time domain plot is simply the instantaneous volt-age plotted versus time. In the frequency domain, the vertical axis is still voltage but the horizontal axis is now frequency. The two representations are consistent but different ways of looking at the same signal. In the frequency domain, the signal has various frequency components (spectral lines) that indicate the amount of energy at each frequency.

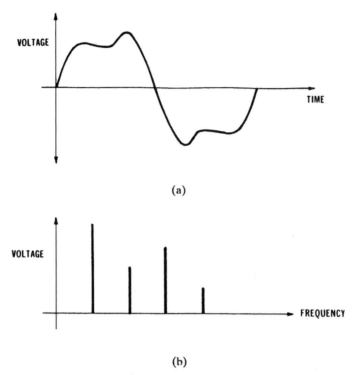

(a)

(b)

Figure 8.1   A signal can be represented in the time domain or in the frequency domain. a) The time domain representation is voltage as a function of time. b) The frequency domain representation is amplitude (voltage) as a function of frequency.

8.2   Bank-of-Filters Spectrum Analyzer

One way to implement a spectrum analyzer is to use a large number of band pass filters, each one tuned to a different frequency. Each filter removes all frequency components of the signal except the particular frequency that the filter was designed to measure. Figure 8.2 shows how each filter picks out a small section of the frequency axis to be measured. At the output of each filter there will be an AC voltage whose amplitude corresponds to the amount of energy contained within the filter's bandwidth. The outputs of each of these filters can then be detected and displayed to produce the frequency domain information (Figure 8.3). This type of instrument is called a *bank-of-filters spectrum analyzer.*

8.2.1   Frequency Resolution

The frequency resolution of the bank-of-filters analyzer is determined by the bandwidth of the individual filters. Figure 8.2 shows that both filter 1 and filter 2 will detect unique spectral lines. The frequency of each of those lines is known only to the extent that

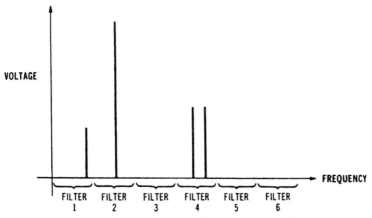

Figure 8.2 A collection of band pass filters can be used to measure the frequency content of a signal. Each filter measures the energy in a small frequency band.

they are within the frequency range of their respective filter. If the filter is 100 Hz wide, then the frequency of the spectral line has an uncertainty of 100 Hz. Another situation is shown in filter 4 where two spectral lines are shown within the same filter's frequency range. These two spectral lines are not detected as being distinct frequency components. Instead, the filter will measure the amount of energy present within its frequency range without regard to how many spectral lines produced that energy. Thus, the ability to resolve two closely spaced spectral lines also depends on the width of the filters.

The bank-of-filters analyzer is simple in concept and results in an instrument capable of quickly tracking changes in the signal's spectral content. The major disadvantage of this approach is that a very large number of narrow filters is needed for most applications. The number of filters required increases with the frequency range and frequency resolution of the analyzer. For example, an analyzer designed to cover 0 to 1 MHz with 1 kHz-wide filters would require 1,000 filters. This limitation prevents the technique from being used for wide-bandwidth instruments. However, this technique is used in low-frequency applications (such as audio and sound level measurements) where a more reasonable number of filters is sufficient.

Although the bank-of-filters technique is not often used in general electronic measurement, it does provide a good conceptual basis for understanding the instrumentation techniques that are commonly used.

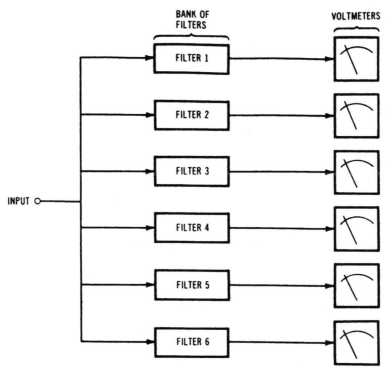

Figure 8.3   The bank-of-filters analyzer. The output of each filter is detected and displayed to provide the frequency domain characteristics of a signal.

8.3   FFT Spectrum Analyzers

FFT spectrum analyzers, also called *dynamic signal analyzers,* use the *Fast Fourier Transform* (or *FFT)* to measure the frequency spectrum of a signal. The FFT is a very efficient mathematical technique for computing the spectrum of a waveform from its time domain representation. First, the analog waveform must be sampled and turned into digital form by means of an analog-to-digital converter (ADC). Then, the FFT algorithm is used to compute the frequency domain representation from the time domain data, which results in the spectrum that can be shown graphically.

8.3.1   Sampling

The input waveform must be turned into digital form before the FFT computation can be applied to it. The waveform is sampled at regular intervals and the voltage at each point is converted into a digital value. Figure 8.4 shows a sine wave and the sampled data points taken from it. Notice that since the points are closely spaced on the sine wave, the waveform can easily be reconstructed by filling in between the sample points. The rate at which the sampling occurs is called *sample rate* and is usually expressed in units of samples/sec or Hertz.

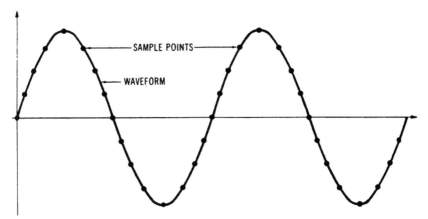

Figure 8.4   The input waveform must be sampled in order for the FFT to be computed.

There must be enough samples taken on each cycle of the waveform so that the frequency spectrum of the waveform can be extracted from the sample points. How fast do we need to sample a given signal? As discussed in Section 4.4, the *Sampling Theorem* states that all of the information in a signal is preserved if it is sampled at a rate that is over twice the highest frequency present. For a spectrum analyzer having bandwidth *BW* and sample rate f_s, the following equation must be true:

$$f_S > 2 \cdot BW$$

If the sample rate is not high enough, *aliasing* can occur. Figure 8.5 shows two waveforms that share a common set of sample points. That is, the sample points fall on both of

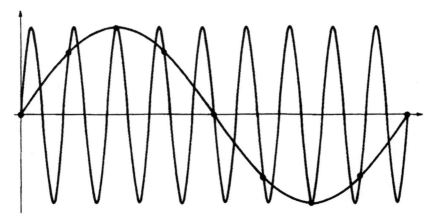

Figure 8.5   Aliasing can occur if precautions are not taken. Shown here are two different waveforms that will result in the same set of sample points. Therefore, after sampling, the two waveforms are indistinguishable.

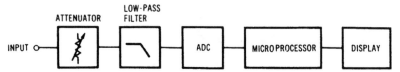

Figure 8.6   The simplified block diagram of an FFT spectrum analyzer has a low-pass filter to prevent aliasing.

these waveforms such that both waveforms produce the exact same sample points. Therefore, the two waveforms cannot be distinguished after sampling has occurred. This phenomenon is called aliasing, since one waveform acts as the alias (or impostor) of the other. This aliasing is highly undesirable in an instrument that is intended to measure the frequency content of a signal. The Sampling Theorem is satisfied for the lower frequency waveform of Figure 8.5, but not for the higher frequency signal.

The common solution for the prevention of aliasing is to filter out any frequencies that would violate the Sampling Theorem. In Figure 8.5, one waveform has a much higher frequency than the other. A low-pass filter is inserted in the signal path to eliminate frequencies greater than half the sample rate, to ensure that the Sampling Theorem is satisfied.[1] The resulting block diagram for the FFT spectrum analyzer is shown in Figure 8.6. The signal being measured passes through the input amplifier and is low-pass filtered to remove any potential alias signals. The waveform is then sampled and converted to digital form by the ADC. The microprocessor receives the digital information and performs the FFT calculation, resulting in the frequency spectrum of the signal.

8.3.2   FFT Computation

The FFT computation is made up of a very complex set of mathematical operations. A stream of digital samples from the time domain are collected and processed as one array of data. The output of the FFT is also an array of data, but in this case it is frequency domain data. The FFT operates as a time domain to frequency domain converter, taking a slice of time domain data and converting it into its corresponding frequency spectrum (Figure 8.7). The output of the FFT computation is equivalent to the output of a bank-of-filters analyzer. Each frequency domain data point corresponds to the output of one of the filters in the bank-of-filters analyzer. The FFT does not require a large number of filters since the same effect is accomplished mathematically.

A typical FFT analyzer might have 1,000 such data points, which means that 1,000 filters would be required in a bank-of-filters analyzer to achieve the same frequency resolution. Most FFT analyzers have the ability to vary the frequency span and hence their

---

1.   As discussed in Chapter 5, many oscilloscopes use the FFT to compute the spectrum of a signal. However, many oscilloscopes do not provide protection against aliasing.

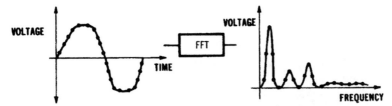

Figure 8.7 The FFT computation takes a slice of the sampled time domain data and computes the frequency domain from it.

frequency resolution (with a constant number of frequency data points). Thus, the bank-of-filters analyzer would need considerably more than 1,000 filters to achieve the same overall performance as an FFT analyzer. Very narrow filter bandwidths are difficult to build using the bank-of-filters approach. However, the FFT computation can produce the equivalent of extremely narrow filters, with bandwidths much smaller than 1 Hz.

The major limitation of the FFT analyzer is the ADC, since the bandwidth of the spectrum analyzer is limited to less than half the sample rate of the ADC. Existing high-resolution ADC technology tends to limit the bandwidth of the typical FFT spectrum analyzer to around 10 MHz. ADC technology will continue to improve, providing faster sample rates and allowing wider bandwidth FFT analyzers to be developed. Another limitation is the computing power of the microprocessor or other digital circuitry used to compute the FFT. The amount of time it takes to compute the FFT may be significant, limiting the speed with which the spectrum can be displayed. Digital technology is improving at a fast rate, enabling dramatic increases in FFT computation speeds.

8.3.3 Specifications

The specifications of a typical FFT spectrum analyzer are shown in Table 8.1. Figure 8.8 shows a spectrum analyzer that uses FFT techniques to characterize signals in the frequency domain.

Table 8.1 Abbreviated Specifications of an FFT Spectrum Analyzer

Specification	Value
Frequency range	0 to 102.4 kHz
Number of channels	2
Measurement range	+27 to –120 dBV
Dynamic range	90 dB
Amplitude accuracy	± 0.15 dB

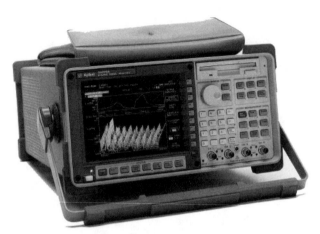

Figure 8.8 A spectrum analyzer that uses the Fast Fourier Transform to compute the spectrum of a signal. (Photo: Agilent Technologies. Reprinted with permission.)

8.4 Wavemeters

The bank-of-filters spectrum analyzer uses many filters, each one tuned to a different frequency, to measure the frequency components present in a waveform. The *wavemeter* accomplishes this by having one filter whose frequency is tunable over the frequency range of interest (Figure 8.9). The tunable filter is tuned to the frequency of interest and the output level of the filter is measured and displayed by a meter. The meter can display only one frequency component at a time, namely the frequency to which the filter is tuned. This is obviously a disadvantage compared to the FFT and bank-of-filters spectrum analyzers, but is acceptable for many applications.

Generally, it is very difficult to design and build filters that can be tuned over a wide frequency range while maintaining a constant filter shape. Thus, the frequency-tunable filter is not usually implemented this way.

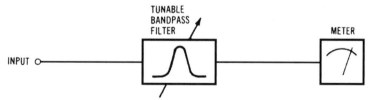

Figure 8.9 The conceptual block diagram for the wavemeter. The wavemeter uses a frequency-tunable filter to measure the signal level at a single frequency.

8.4.1 The Practical Wavemeter

Instead of tuning the filter over a frequency range, it is easier to leave the filter at a fixed frequency and move the signal in frequency. This technique is shown in Figure 8.10. Starting from the right side of the figure, the filter is designed to measure the level at the *intermediate frequency* (IF). The *mixer* and *local oscillator* (LO) shift the input signal in frequency to the intermediate frequency. Thus, the filter can be at a fixed frequency at the expense of shifting the input signal in frequency.

A mixer takes the input frequency (f_{IN}) and the local oscillator frequency (f_{LO}) and produces the sum and difference frequencies:

$$f_{LO} - f_{IN}$$

$$f_{LO} + f_{IN}$$

For every input frequency, there are two frequencies at the output of the mixer.[2] One of these frequencies is ignored by the IF filter and the other is passed through and measured. The local oscillator's frequency is variable and is adjusted by the user to cause the mixing process to produce the desired frequency at the IF filter. The function of the low-pass filter will be discussed later.

This method of frequency translation is known as the *superheterodyne technique* and is the same technique used in most modern radio receivers. A numerical example will help clarify this mixing process. Figure 8.11 shows a wavemeter designed to measure from 0 to

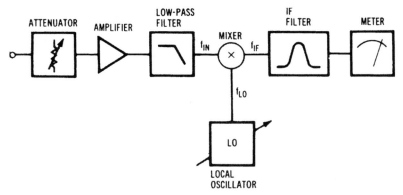

Figure 8.10 A more practical block diagram for the wavemeter uses a mixer and local oscillator to shift the input signal in frequency. The intermediate frequency filter is tuned to the output of the mixer.

---

2. Actually there will be many different frequencies present at the output, but for the purposes of understanding the block diagram, only two are significant.

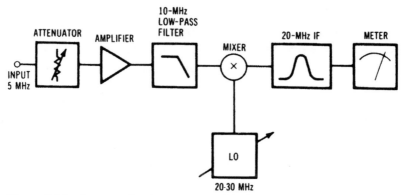

Figure 8.11   An example of a wavemeter block diagram with frequencies specified.

10 MHz. The LO operates from 20 to 30 MHz and the IF filter is tuned to 20 MHz. Suppose the wavemeter was tuned to measure a signal at 5 MHz. The LO is tuned to 25 MHz, resulting in 25 ± 5 MHz (equals 20 MHz and 30 MHz) out of the mixer. The 20-MHz signal falls directly on the IF filter and is measured while the 30-MHz signal is ignored. If the wavemeter were tuned to 6 MHz, then the LO would be tuned to 26 MHz and the mixer output frequencies would be 20 MHz and 32 MHz. The instrument controls are, of course, designed such that the LO frequency adjustment indicates the wavemeter frequency and not the frequency of the LO.

So what about the low-pass filter (also known as the *image filter*)? Again, referring to Figure 8.11, assume that the wavemeter is tuned to 5 MHz (with the LO at 25 Hz). If a 45-MHz signal were present at the input to the mixer, it would mix with the 25-MHz LO frequency and produce sum and difference frequencies (20 and 70 MHz). The 70-MHz signal would not cause any problems but the 20-MHz signal would fall directly on the IF. Therefore, the wavemeter would not be able to distinguish between a 5-MHz signal and a 45-MHz signal. The undesired signal at 45 MHz is known as an *image*. Image signals occur at frequencies equal to the LO frequency plus the IF frequency (for the block diagram shown in Figure 8.11). Or equivalently, the image is at the input frequency plus two times the intermediate frequency.

As a practical matter, high performance wavemeters are much more complex than the single IF block diagram shown. Usually several sets of intermediate frequency filters and mixers are used to implement a wavemeter. Conceptually, these block diagrams use the same frequency translation techniques to move the signal past the filter to eliminate the need for a variable frequency filter. Although practical wavemeters usually do not use variable frequency filters, the concept is still a valuable tool for understanding the function performed by a wavemeter since the superheterodyne block diagram produces the same effect as the variable frequency filter approach.

8.4.2 Measurement Example

Suppose the frequency components of a 1-MHz square wave are to be measured. Figure 8.12 shows the time domain representation and the frequency domain points measured by a wavemeter. Tuning the wavemeter to 1 MHz allows the amplitude of the fundamental to be measured, while ignoring the harmonics. Similarly, when the wavemeter is tuned to 3 MHz and then 5 MHz, the third and fifth harmonics (respectively) are measured (while ignoring the other frequency components). Other frequencies can be measured as needed, but the wavemeter is a "one frequency at a time" kind of instrument. Note that a simple voltmeter cannot make this measurement since it is inherently a broadband instrument without controlled frequency selectivity. Performing this type of measurement confirms that periodic signals really do have harmonics in the frequency domain.

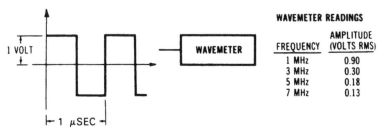

WAVEMETER READINGS

FREQUENCY	AMPLITUDE (VOLTS RMS)
1 MHz	0.90
3 MHz	0.30
5 MHz	0.18
7 MHz	0.13

Figure 8.12 An example of a wavemeter measurement. The wavemeter is used here to measure the individual frequency components of the square wave.

8.5 Resolution Bandwidth

The bandwidth of the IF filter determines the frequency resolution of the instrument (similar to the bank-of-filters spectrum analyzer). Thus, the bandwidth of the IF filter is often referred to the *resolution bandwidth*. If more than one IF filter is used in an instrument, the narrowest one dominates and is considered the resolution bandwidth. (Don't confuse the resolution or IF bandwidth with the bandwidth of the instrument, which refers to the overall frequency range of the instrument.)

A wavemeter may have selectable resolution bandwidths to provide for some flexibility in how the measurement is made. The choice of resolution bandwidth depends on several factors. Filters take some amount of time to settle. That is, when a signal first appears at the input of a filter, it will take some time before the signal appears at the output. In addition, the output of the filter will take some time to settle to the correct value, so that it can be measured. When the signal first appears at the output, it varies in amplitude and later settles out to a constant value. The narrower the filter bandwidth, the longer the settling time. In many cases, this time is so small that it is inconsequential. But if very narrow bandwidths are used (which provide better frequency resolution), the settling time can become significant.

Narrow bandwidths also increase the difficulty of tuning to a particular signal, particularly if the signal frequency is unknown or varies. The wavemeter must be tuned exactly to the signal frequency, otherwise the signal will fall outside the resolution bandwidth. If the signal or wavemeter frequency drifts, then maintaining the same frequency is more difficult. Using a wider resolution bandwidth makes the frequency difference between the signal and wavemeter less critical. Compounding the problem is the fact that narrower bandwidths take longer to settle. For example, a 10 Hz-wide filter might take on the order of a second to settle to an accurate value. If an unknown frequency is being measured, it takes several (if not many) adjustments to tune the wavemeter to the signal frequency. So as narrower bandwidths are used, the settling and adjustment times increase while the required accuracy of the frequency adjustment also increases.

The choice of resolution bandwidth will depend on the signal being measured. Figure 8.13 shows two signals very close together in frequency. If the two spectral lines are to be measured individually, then a narrow bandwidth is required (as shown). If a wider bandwidth is used, then the energy of both signals will be included in the measurement. (This could be desirable in some cases.)

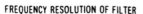

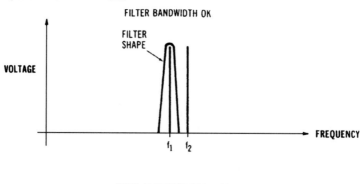

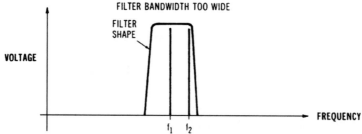

Figure 8.13   In order to distinguish between two closely spaced signals, a narrow bandwidth is required. If a wide bandwidth is used, then both signals will be included in the measurement.

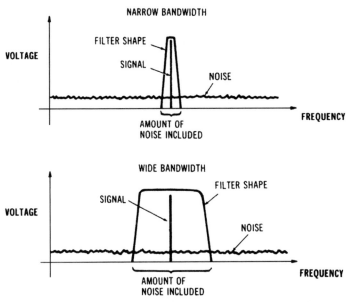

Figure 8.14 Narrowing the resolution bandwidth of the instrument results in a smaller amount of noise corrupting the measurement.

There is always some amount of noise present in a measurement. For large signal levels, noise is often so small it can be ignored, but for low-level measurements, noise will become a significant contributor to measurement error. Noise is generally broadband in nature, existing across a broad range of frequencies. With significant noise included in the measurement, the measured value will be in error (too large) depending on the noise level. Figure 8.14 shows a signal being measured in the presence of noise. With a wide bandwidth, more noise is included in the measurement. With a narrow bandwidth, very little noise enters the resolution bandwidth filter and the measurement is more accurate. The effect of noise on a voltage measurement is proportional to the square root of the resolution bandwidth.

8.6   Narrowband and Broadband Measurements

The bank-of-filters analyzer, the FFT analyzer, and the wavemeter are all examples of narrowband measurements. Although the first two instruments measure a range of frequencies simultaneously, the techniques they use are the same as a narrowband measurement. The wavemeter is more obviously narrowband because it acts like a variable frequency filter tuned to the frequency of interest. On the other hand, instruments such as voltmeters and oscilloscopes are broadband. That is, they do not have the ability to look at frequencies selectively, but instead must measure across their entire frequency range at all times.

As an example, consider the frequency domain representation of the sine wave shown in Figure 8.15a. Ideally, the sine wave is a single, infinitely thin spectral line. In reality, some or all of the imperfections shown in Figure 8.15b (harmonics, spurious responses, and noise) may be present. A broadband measuring instrument such as a voltmeter would include the fundamental, harmonics, spurious responses, and noise in the measurement. This may be desirable if the goal of the measurement is to determine the total signal level present across a wide bandwidth. In many other cases, it is desirable to ignore the imperfections, especially the noise and spurious responses that are not part of the signal. (Is the harmonic part of the signal? Sometimes.) A narrowband measuring instrument such as a wavemeter will filter out all but the desired frequency components, resulting in a measurement that includes only the fundamental.

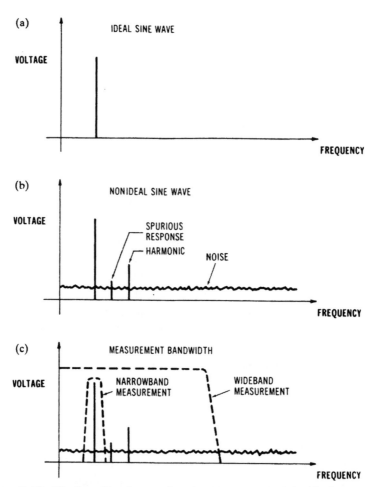

Figure 8.15   The benefits of narrowband measurement. a) An ideal sine wave is a single spectral line. b) The sine wave may be accompanied by a variety of imperfections. c) The broadband measurement includes everything in the measurement. A narrowband measurement can select the desired frequency component.

Suppose only the harmonics or only the spurious responses were to be measured. The voltmeter is useless, but the wavemeter could be tuned to the proper frequency to make the measurement. (The bank-of-filters and FFT spectrum analyzers could also be used to make the measurement. In fact, they would be more convenient since they measure the entire frequency range at one time.) What about the noise level—could it be measured in the presence of a large signal? Again, the broadband voltmeter would be useless since it includes everything in the measurement. A narrowband instrument, however, could be tuned to the frequency of interest and (assuming no spectral lines happened to be at that frequency) could measure the noise level there. The bandwidth of the measurement would have to be taken into account since the resolution bandwidth will affect how much noise is measured.

8.7 Swept Spectrum Analyzers

The wavemeter block diagram can be improved one step further. The wavemeter measures at only one frequency at a time, but if the wavemeter could be tuned or swept across its frequency range, then the entire spectrum of a signal could be automatically characterized. The *swept spectrum analyzer* is essentially a wavemeter that automatically sweeps in frequency and displays the results. Like the bank-of-filters and the FFT spectrum analyzers, the result is a frequency domain display of the signal. Figure 8.16 shows the concept of sweeping the wavemeter, which acts like a variable frequency filter.

Figure 8.17 shows a conceptual block diagram of a swept spectrum analyzer. *A voltage-controlled oscillator* (VCO) is used as the local oscillator to allow the frequency of the analyzer to be changed easily. A ramp generator drives the tuning voltage of the VCO, causing it to sweep up in frequency repetitively. The same ramp voltage drives the horizontal axis of the display. The image filter, mixer, and IF filter all function the same as they did with the wavemeter. The output of the IF filter, however, goes to the *detector,* which detects the sine wave, producing a DC level that is proportional to the level present in the IF filter. This signal drives the vertical axis of the display, resulting in the detected IF level being painted across the display while the VCO sweeps. The result is an amplitude versus frequency display of the signal.

VOLTAGE

FREQUENCY

Figure 8.16 A swept spectrum analyzer can be thought of as a wavemeter that is swept in frequency. The result is a filter that automatically sweeps in frequency, characterizing the signal at its input.

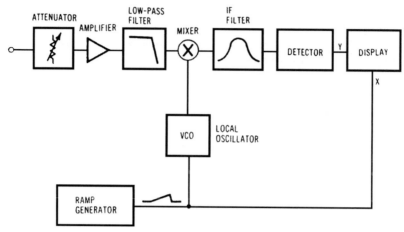

Figure 8.17   The conceptual diagram of the swept spectrum analyzer. The local oscillator is swept in frequency while the output of the intermediate frequency filter is displayed on the screen.

The operation of the analyzer has been described in an analog sense. Digital technology has replaced many of the analog circuits. In particular, the output of the detector may be measured by an analog-to-digital converter, which is connected to a microprocessor. In many cases, the ADC has been moved into the IF section and the resolution bandwidth filtering and detection are done digitally. The microprocessor receives the data in digital form and processes it for display. The input amplifier, image filter, mixer, and VCO are usually implemented using analog technology. This allows the analyzer to operate at radio and microwave frequencies, which may not be possible with digital techniques. As digital technology improves, its use will migrate forward in the block diagram toward the input amplifier.

8.7.1   Sweep Limitations

The user normally has control over the *sweep time* (the elapsed time of each sweep, sometimes called the *scan time*), the frequency range over which the analyzer sweeps and the resolution bandwidth. The analyzer cannot be swept arbitrarily fast while maintaining its specified accuracy, but will have a sweep rate limitation depending on the resolution bandwidth chosen. The sweep rate (expressed in Hertz per second) is not usually chosen directly by the user, but is determined by the frequency range swept divided by the sweep time.

The limitation on sweep rate comes from the settling or response time of the resolution bandwidth filter discussed earlier. If an analyzer is swept very quickly, the filter does not have time to respond and the measurement is inaccurate. Under such conditions, the analyzer display tends to have a "smeared" look to it, with spectral lines being wider than normal and shifted to the right. In general, the maximum sweep rate of an analyzer is given by:

$$sweep\ rate\ =\ \frac{BW^2}{k}$$

where BW is the bandwidth of the resolution filter and k is a factor depending on the resolution bandwidth filter shape (typically k is equal to 2). Notice that the sweep rate is proportional to the bandwidth squared. There may be other limitations on sweep rate in the instrument, such as the speed at which the local oscillator can sweep.

Fortunately, most instrument manufacturers have designed in mechanisms that unburden the user from having to worry about the sweep rate. In older instruments, this is usually accomplished by some sort of mechanical interlock and/or a warning light that tells the user that the measurement may be uncalibrated. In modern instruments, microprocessor software chooses the fastest accurate sweep time. In both cases, an informed user is protected from making an erroneous measurement. At the same time, the user is given the option of over-riding the built-in sweep rate protection. There may be valid reasons for doing this, but the user must proceed at the risk of making an inaccurate measurement.

Since the sweep rate is proportional to the square of the resolution bandwidth, decreasing the resolution bandwidth dramatically decreases the maximum sweep rate. For a given swept frequency range, the sweep time is inversely proportional to the square of the resolution bandwidth. Thus, the user will find that the sweep time (and, therefore, the measurement time) must be significantly increased when narrow resolution bandwidths are used. FFT techniques result in much narrower resolution bandwidths for a given measurement time; or they give the same resolution bandwidth much faster.

8.7.2   Dynamic Range

The *dynamic range* of a spectrum analyzer is the difference between the largest signal and the smallest signal that can be reliably measured *at the same time* (Figure 8.18). Compare this to *measurement range,* which is the difference between the largest signal that can be measured and the smallest signal that can be measured, not simultaneously. The smallest signal that can be measured is limited by noise, distortion, and spurious responses present in the analyzer.

For low-level spectrum analyzer measurements, even a small amount of noise may be significant. When a signal falls below the noise level present in the instrument, that signal can no longer be reliably measured. The amount of internal noise in the spectrum analyzer is usually specified in terms of *displayed average noise level (DANL).* DANL is the average noise level shown on the spectrum analyzer display with the input terminated with a Z_0 load, normalized to a 1-Hz bandwidth.

Spectrum analyzers also produce a certain amount of distortion in their measurement circuits. This means that a perfectly clean signal may be displayed as having some small harmonics or other spectral impurities due to the distortion in the instrument. When a signal

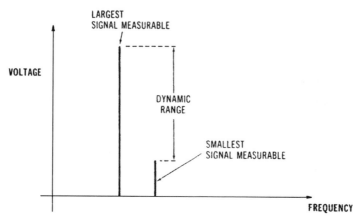

Figure 8.18 The dynamic range of a spectrum analyzer is the difference between the largest signal and the smallest signal measurable at the same time. The smallest signal measurable is limited by noise, distortion, and residual responses.

falls below these distortion products present in the analyzer, the user cannot distinguish between a valid signal and an erroneous one. In addition, some residual spurious responses may be present in the instrument. Distortion products will disappear when the input signal is removed, but residual responses will be present with or without an input signal. All of these imperfections in the instrument are specified by the manufacturer, either separately or in terms of dynamic range.

8.7.3   Effect of Resolution Bandwidth

As previously mentioned, using a narrower resolution bandwidth causes less noise to be present in the measurement. The amount of noise in the circuit remains the same, but a smaller amount of it is measured due to the reduced bandwidth. Figure 8.19a shows a typical spectrum measurement having a signal present above the broadband noise "floor." Note that the noise is generally present at all frequencies and may be present in the signal being measured or may be the internal noise of the analyzer. If the resolution bandwidth is narrower, the noise floor drops on the display (Figure 8.19b). Again, this is because the IF filter of the analyzer has been made narrower in bandwidth, which lets in less noise. A factor of 2 change in resolution bandwidth causes a 3-dB change in the noise level. As the measured noise level drops (due to narrowing the bandwidth), smaller signals that were previously obscured by the noise can be measured. This has the effect of increasing the dynamic range of the measurement at the expense of increased sweep time.

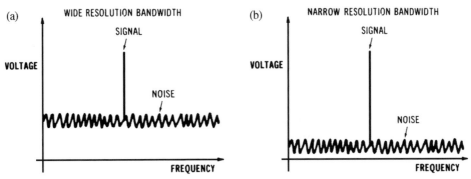

Figure 8.19 The effect of changing the resolution bandwidth. a) A wider resolution bandwidth has a higher noise level on the display. b) Narrowing the resolution bandwidth causes the noise level to drop.

8.7.4   Effect of Video Bandwidth

Most analyzers include another type of filtering after the detector called *video filtering*. This filter also affects the noise on the display, but in a different manner than the resolution bandwidth. Figure 8.20 shows the positioning of the video filter in the swept spectrum analyzer block diagram. Note that the resolution bandwidth filter is in front of the detector (*predetection filtering*) and the video filter is after the detector (*postdetection filtering*). The effect of video filtering is shown in Figure 8.21. Figure 8.21a shows a signal being measured in the presence of noise with a wide video filter. Figure 8.21b shows the same signal and noise with a narrow video filter. The average level of the noise remains the same but the variation in the noise is reduced. The effect on the analyzer's display is that the noise floor compresses into a thinner trace, while the position of the trace remains the same. (Compare this with the effect of the resolution bandwidth, which reduces the level of the noise.)

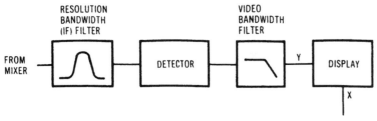

Figure 8.20 A portion of the swept analyzer block diagram showing the position of the video bandwidth filter.

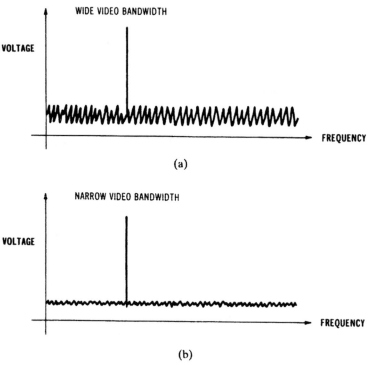

Figure 8.21 The effect of changing the video bandwidth. a) Signal in the presence of noise with a wide video bandwidth. b) The same signal and noise with a narrow video bandwidth. The average noise level remains the same, but the variation in the noise is reduced.

8.7.5 Tracking Generator

Many swept spectrum analyzers include a sine wave output that tracks the analyzer's input frequency as it sweeps. Since this sine wave is always at the same frequency that the analyzer is measuring, it allows convenient frequency response characterization of circuits. This *tracking generator* output is connected to the input of the circuit being tested and the output of the circuit is connected to the input of the analyzer (Figure 8.22). The resulting frequency response is plotted on the spectrum analyzer display. Recall from Chapter 5 that characterizing the frequency response of a circuit (such as a filter or amplifier) can be done on a point-by-point basis using a source and a scope. A spectrum analyzer with a tracking generator can accomplish the same thing, but much more quickly.

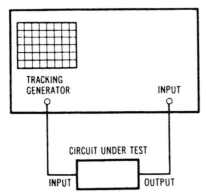

Figure 8.22 A spectrum analyzer having a tracking generator output can be used to measure the frequency response of a circuit.

8.7.6 Spectrum Analyzer Inputs

At frequencies below about 50 MHz, spectrum analyzers are usually provided with both high impedance (1 MΩ oscilloscope type inputs) and 50-Ω inputs. The high impedance input often has degraded specifications associated with it and the 50-Ω input is the high-quality input. As the upper limit of the frequency range increases, the high impedance input becomes impractical due to the effect of the input capacitance and only the 50-Ω input is included. Although not as common as 50-Ω inputs, 75-Ω inputs are also found on some spectrum analyzers. Lower frequency FFT instruments usually are supplied with only high impedance inputs.

8.7.7 Advanced Features

Modern spectrum analyzers are microprocessor controlled, which allows more advanced features to be implemented. A *marker* (or *cursor*) is supplied, which can be located at any point on the displayed trace, with the frequency and amplitude values at that point displayed digitally on the display. A *marker reference* or *offset* can be specified so that the marker reads relative to the reference point. The marker can usually be configured to display whatever units are convenient to the user (dBm, dBV, volts, etc.). Other marker features allow the user to assign the current marker frequency to the center frequency and the current marker value to the full-scale value (reference level).

Digital storage of the displayed trace keeps the display from flickering at the sweep rate. Additional trace storage may be provided, as well as the ability to display the difference between the current and stored traces. Other features provided include automatic noise measurement, a built-in frequency counter, and manual sweep (not a sweep at all, but a single frequency measurement).

A precision RF microwave spectrum analyzer is shown in Figure 8.23 and the specifications of a typical swept spectrum analyzer are shown in Table 8.2. A very important difference

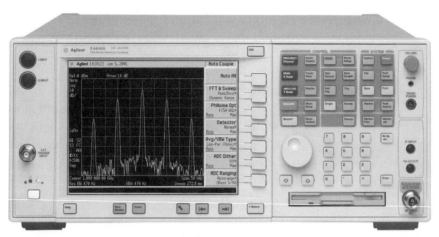

Figure 8.23   A high-performance RF/microwave spectrum analyzer. (Photo: Agilent Technologies. Reprinted with permission.)

between the FFT analyzer and the swept analyzer is the frequency range that is covered, with the swept analyzer having a much higher frequency range. It should be pointed out that the typical specifications shown are for a typical high-frequency analyzer—commercially available microwave spectrum analyzers have frequency ranges that extend beyond 40 GHz.

Table 8.2   Abbreviated Specifications of a Swept Spectrum Analyzer

Specification	Value
Frequency range	3 Hz to 6.7 GHz
Measurement range	+25 to –150 dBm
Displayed average noise level (DANL)	–150 dBm (preamp off)
	–168 dBm (preamp on)
Dynamic range	80 dB
Amplitude accuracy	± 0.65 dB
Frequency sweep time	1 ms to 2,000 sec
Resolution bandwidth	1 Hz to 8 MHz
	(10% steps)

8.7.8   Combined FFT and Swept Analyzer

As more of the swept spectrum analyzer is implemented using digital technology, there is a blurring between the block diagram of the FFT analyzer and the swept analyzer. By applying an FFT to the IF of the swept analyzer, we can get the best of both techniques. The front end of the swept analyzer provides the wide frequency coverage normally associated with swept analyzers. At the same time, using the FFT on the IF signal provides narrow resolution bandwidth measurements without the sweep rate limitations of the swept analyzer. The net result is that narrowband measurements can be made quickly at very high frequencies.

8.8 Spectrum Analyzer Measurements

Regardless of the implementation, the spectrum analyzer presents the instrument user with a frequency domain view of a signal, just as the oscilloscope provides a picture of the time domain. A variety of measurements can be made using the spectrum analyzer that simply involve interpreting the frequency domain display of a signal. The simplest type of spectrum analyzer measurement is just measuring the amplitude and frequency of a single sine wave. Although analyzers are well suited for such measurements, the real power of the spectrum analyzer is utilized when more complex spectra are measured.

Harmonic distortion is often present in signals, as shown in Figure 8.24a. Ideally, a sine wave would be a single spectral line at the fundamental frequency. With harmonic distortion present, some of the waveform's energy appears at the harmonic frequencies. Note that the spectral lines are at multiples of the fundamental frequency. A good spectrum analyzer is capable of measuring distortion products that are 80 dB smaller than the fundamental (0.01% distortion), which is equivalent to a factor of 10,000 in voltage. Suppose, for example, that amount of distortion is present in a 1-volt RMS sine wave. The corresponding harmonic distortion level is 1/10,000 = 100 μV. Harmonic distortion that small would not be detectable by the most sensitive oscilloscope (because the scope must display the rather large fundamental waveform simultaneously). The best oscilloscope displays are capable of resolving distortion in the range of a few percent for a trained operator. Thus, the spectrum analyzer is far superior than an oscilloscope for making distortion measurements. Harmonic distortion is generally measured with a spectrum analyzer in decibels relative to the fundamental, but can also be expressed in percent distortion.

Intermodulation distortion occurs when two sine waves are present at the same time in a circuit that exhibits distortion. Ideally, the addition of two sine waves simply results in two unique spectral lines. If distortion is present in the system, the two signals will intermodulate. In addition to producing the harmonics of both waveforms, this causes several types of sum and difference frequencies to be generated. For instance, as shown in Figure 8.24b, it is common for spectral lines to appear at the original frequencies plus or minus the difference between the two original frequencies.

Modulation sidebands may appear on a signal due to intentional or unintentional amplitude modulation (AM), frequency modulation (FM), or phase modulation (PM). Modulation is intended to be on many communication signals, including standard AM and FM radio broadcast signals. In other cases, the modulation may be a by-product of some other circuit operation. In either case, it is important to be able to characterize the signal accurately. The spectrum analyzer inherently measures the absolute sideband level, which may be expressed

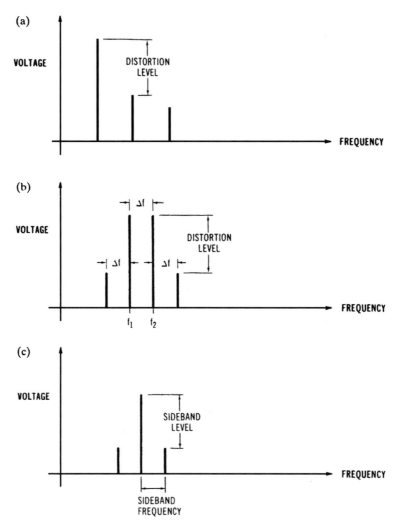

Figure 8.24 The most common spectrum analyzer measurements involve interpretation of the frequency domain display provided by the analyzer. a) Harmonic distortion measurement. b) Intermodulation distortion measurement. c) Measurement of modulation sidebands.

in decibels relative to the carrier frequency. Depending on the type of modulation, this relative sideband level can be related back to percent modulation, frequency deviation, and so on.

Figures 8.25 through 8.28 show some actual spectrum analyzer measurements of some typical signals. A relatively pure sine wave is shown in Figure 8.25, with a start frequency (left side of display) of 0 Hz and a stop frequency (right side of display) of 50 MHz. With 10 divisions across the display, this means each division represents 5 MHz. The frequency of the sine wave is 5 MHz and the third harmonic is just visible above the noise floor at 15 MHz (three divisions from the left). The fifth and seventh harmonics are also visible but are in danger of being confused with the peaks of the noise floor.

The spectrum of a 5-MHz square wave has very strong odd harmonics (Figure 8.26). Here, the fundamental, third harmonic, fifth harmonic, seventh harmonic, and ninth harmonic are all easily visible, well above the noise floor.

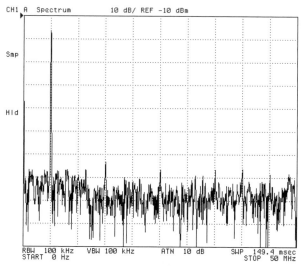

Figure 8.25 The spectrum analyzer measurement of a relatively pure 5-MHz sine wave shows low-level harmonics just above the noise floor of the measurement.

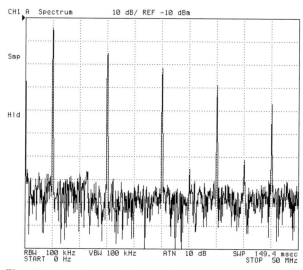

Figure 8.26 The spectrum analyzer measurement of a square wave shows the presence of strong odd harmonics.

Amplitude modulation causes sidebands on both sides of the carrier (Figure 8.27). In this case, the carrier frequency is 5 MHz and the modulating frequency is 10 kHz. Note that the spectrum analyzer is centered on 5 MHz with a frequency span of 100 kHz. The vertical scale is 10 dB/division, so the sidebands are roughly 20 dB below the carrier amplitude.

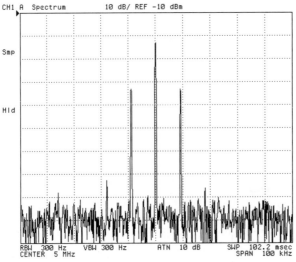

Figure 8.27   The spectrum analyzer measurement of an amplitude-modulated signal has sidebands on both sides of the carrier.

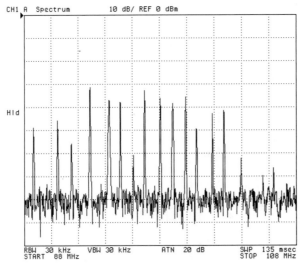

Figure 8.28   An antenna attached to the input of a spectrum analyzer captures local FM broadcast radio stations in the range 88 MHz to 108 MHz.

The spectral content of a spectrum measurement can be very complex. Figure 8.28 shows a measurement that is made "off the air" by connecting a small telescoping antenna to the input of the analyzer. The frequency span is set to cover 88 MHz to 108 MHz, which covers the FM broadcast band in the United States. More than 15 different FM radio stations are shown in the measurement.

8.9 Network Analyzers

In the preceding chapters, we've seen multiple ways of measuring the characteristics of a network or circuit. In Chapter 5, we saw how a source and oscilloscope can be used to make gain and phase measurements on a circuit. As discussed in this chapter, a spectrum analyzer with a tracking generator can be used to measure the frequency response of a circuit. These are fundamental measurements used during the design and manufacture of circuit components.

A *network analyzer* is an instrument that automatically measures the characteristics of a circuit over a selectable frequency range. A network analyzer may have a sine wave source built in or use an external source. The source sweeps in frequency while the input of the analyzer measures the output of the circuit, resulting in a frequency response measurement for the gain through the circuit (Figure 8.29).

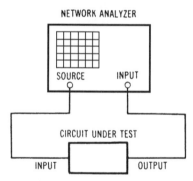

Figure 8.29 A network analyzer provides the stimulus to the device under test and measures the circuit's response.

Network analyzers typically have more than one input channel for increased measurement flexibility and accuracy. Figure 8.30 shows a two-channel network analyzer using a power splitter to supply the source signal simultaneously to the device under test and Channel A of the analyzer. The analyzer measures the ratio of Channel B and Channel A (B/A). Variations in the source level will be measured by Channel A and are effectively removed from the measurement. Without *ratioing*, these source variations would introduce an error in the measurement of the circuit's response.

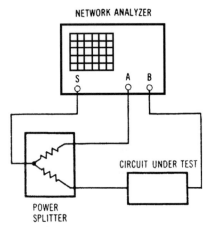

Figure 8.30 A typical measurement connection using a two-channel analyzer and a resistive power splitter. The source supplies a sine wave to both the A input and the circuit under test. The analyzer is configured to measure the ratio of Channel B and Channel A.

Measurements that involve only the magnitude of the signals, such as the gain of a circuit, are called *scalar measurements*. Measurements that include the phase of the signal are called *vector measurements*. Consistent with this terminology, network analyzers may be categorized as *scalar network analyzers* or *vector network analyzers* depending on whether they have phase measurement capability. The signal detection in a network analyzer may be either *broadband* or *narrowband*. Scalar network analyzers usually employ broadband detection, which can usually be accomplished at a lower cost. Vector network analyzers normally use narrowband detection.

Network analyzers are often designed using techniques similar to a swept spectrum analyzer. The use of narrowband IF filters allows a network analyzer to achieve a dynamic range in excess of 100 dB. The conceptual block diagram of a network analyzer with narrowband detection is shown in Figure 8.31. The network analyzer block diagram is similar to the spectrum analyzer's, although there are some practical differences in how they are actually implemented. A network analyzer is usually assumed to be measuring a known frequency generated by the analyzer's source. A spectrum analyzer, on the other hand, must be capable of measuring unknown and arbitrary frequencies. Thus, the network analyzer may not include the same level of image rejection that a spectrum analyzer does. A vector network analyzer must include a means of measuring phase, while this is not required on a spectrum analyzer. Some instruments include both spectrum and network analyzer functions for maximum flexibility. A typical vector network analyzer is shown in Figure 8.32. The specifications of a typical network analyzer are shown in Table 8.3.

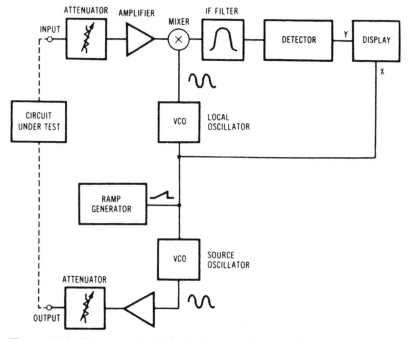

Figure 8.31 The conceptual block diagram of a network analyzer is similar to a spectrum analyzer block diagram except for the addition of a source output and its associated circuitry.

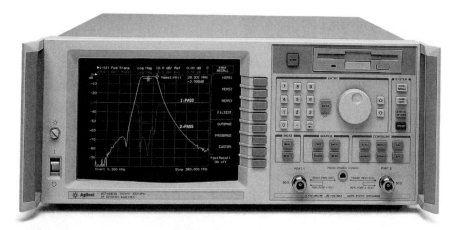

Figure 8.32 A 3-GHz network analyzer with a built-in S-parameter test set. (Photo: Agilent Technologies. Reprinted with permission.)

Table 8.3   Abbreviated Specifications of a Network Analyzer

Specification	Value
Frequency range	300 kHz to 3.0 GHz
Receiver	
Amplitude Accuracy (Transmission)	$\pm$ 0.2 dB
Phase accuracy (Transmission)	$\pm$ 1 degree
Dynamic range	110 dB
Source	
Level accuracy	$\pm$ 1 dB
Output range	+10 dBm to –60 dBm

8.9.1   Two-Port Networks

The measurement shown in Figure 8.30 applies a sine wave to the input of the circuit under test while the output of the circuit is monitored. Since this circuit has two ports (an input and an output), it is referred to as a *two-port network*. To fully characterize a two-port network, we need to measure the input characteristics, the output characteristics, the effect of the input on the output and the effect of the output on the input.

As the frequency increases, transmission line effects affect how signals propagate to and from the circuit under test. With low-frequency measurements and reasonable length cables, we can usually assume that the voltage is the same at both ends of the cable. That is, any signal applied to one end of the cable instantly appears at the other end. At high frequencies, the time it takes for the signal to propagate down the transmission line is significant. In fact, we had better be using a transmission line and not just any old cable to ensure that the signal transmission is in a controlled environment.

8.9.2   S-Parameters

The common way of dealing with the traveling wave behavior in two-port measurements is to use *scattering parameters* (S-parameters). Figure 8.33 shows a two-port network with incident and reflected waves shown at both ports. In general, when a voltage wave, a_1, is incident on port 1, there will be a voltage wave, b_1, reflected back.[3] Similarly, at port 2, an incident wave, a_2, produces a reflected wave, b_2. Since there is usually coupling between the two ports, the incident wave on one port may also produce a wave leaving the other port.

---

3.   This is the frequency domain concept that corresponds to the time domain reflections covered in
       Section 5.15, however, different notation (a_1, b_1, a_2, b_2) is used here, consistent with the common s-
       parameter usage.

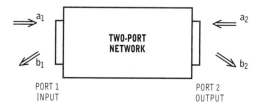

Figure 8.33 A two-port network with incident and reflected waves at both ports.

For a two-port network, four S-parameters are defined that can characterize the behavior of the network.

$$S_{11} = \frac{b_1}{a_1}\bigg|_{a_2 = 0}$$

S_{11} is called the *input reflection coefficient*, and is the ratio of the reflected wave to the incident wave at port 1, with the output properly terminated.

$$S_{21} = \frac{b_2}{a_1}\bigg|_{a_2 = 0}$$

S_{21} is called the *forward transmission coefficient*, and is the ratio of the wave leaving port 2 to the incident wave at port 1, with the output properly terminated. S_{21} is roughly equivalent to the forward gain of the network.

$$S_{22} = \frac{b_2}{a_2}\bigg|_{a_1 = 0}$$

S_{22} is called the *output reflection coefficient*, and is the ratio of the reflected wave to the incident wave at port 2, with the input properly terminated.

$$S_{12} = \frac{b_1}{a_2}\bigg|_{a_1 = 0}$$

S_{12} is called the *reverse transmission coefficient*, and is the ratio of the wave leaving port 1 to the incident wave at port 2, with the input properly terminated. S_{12} is roughly equivalent to the reverse gain of the network.

Scattering parameters are complex numbers that vary with frequency. The magnitude and phase of the S-parameters are often displayed. For example, the magnitude of S_{21} may be displayed on a logarithmic scale to produce the gain of the network in decibels.

S-parameters are a powerful tool for analyzing and measuring radio frequency and microwave circuits and can be somewhat intimidating to the new network analyzer user.

The basics of S-parameters are discussed here to give the reader a cursory understanding of these two-port parameters.[4]

Table 8.4   Quick Reference Guide to S-parameters

S-parameter Hints
S-parameters are normally used at radio or microwave frequencies to characterize networks.
S-parameter measurements are made in a controlled, transmission line environment, typically 50 Ω or 75 Ω
S_{11} and S_{22} are measures of the reflection coefficient of the input and output ports (respectively).
S_{21} and S_{12} are measures of the transfer characteristics of the network. S_{21} corresponds to the forward gain of the network.

8.9.3   Measurement Configurations

Measuring the individual waves that are incident on and reflected from a device requires the use of additional measurement accessories. A *directional coupler* or *bridge* is used to extract the desired wave for measurement. Figure 8.34 shows a three-channel network analyzer configured for measuring the reflection and transmission of a device. The network analyzer source drives a power splitter, which feeds the source signal to the reference channel (R) of the network analyzer and the device under test. A directional coupler or bridge, which extracts the reflected signal from the input port, is inserted in front of the input to the device under test. The output of the device under test is connected to Channel B of the analyzer. The network analyzer measures the reflection coefficient (S_{11}) via the ratio of Channel A to Channel R, so that the source imperfections are removed from the measurement. The transmission through the device (S_{21}) is measured as the ratio of Channel B to Channel R.

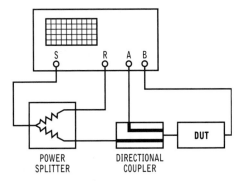

Figure 8.34   A three-channel network analyzer is configured for simultaneous reflection and transmission measurements.

---

4.   For a more thorough mathematical treatment of S-parameters, see the Agilent Technologies Application Note AN 154. For more information on practical S-parameter measurements, see Witte, 2001.

Measurement of all four S-parameters requires additional measurement hardware, including an additional directional coupler or bridge. The required measurement accessories are often configured into one unit, known as an *S-parameter test set*. These test sets have a variety of configurations depending on the specific characteristics of the network analyzer, but they generally include sufficient test accessories to measure all four S-parameters without having to change the connections to the device under test.

8.9.4   Network Analyzer Measurements

The display of a network analyzer provides a plot of the magnitude and/or phase response of a two-port circuit as a function of frequency. The most common format for the display is gain in dB versus frequency. Some examples of basic network analyzer measurements are shown in Figure 8.35. The frequency response plot of a low-pass filter gives the

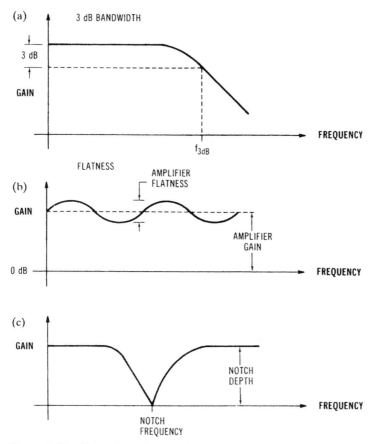

Figure 8.35   Examples of network analyzer measurements. The most common network analyzer display is gain in dB versus frequency. a) Measurement of 3 dB bandwidth. b) Measurement of amplifier gain and flatness. c) Measurement of filter notch depth and frequency.

complete response of the filter as well as allows measurement of the 3-dB bandwidth (Figure 8.35a). The characteristics of an amplifier can also be measured (Figure 8.35b). In addition to the nominal gain of the circuit, the flatness of the response can easily be found. Rejection measurements, requiring a wide dynamic range, are common in electronic filter work (Figure 8.35c). The depth and frequency of a notch filter are measured.

Several measurements of a band-pass filter are shown in Figures 8.36 through 8.38. The gain of the filter, in log magnitude format, is shown in Figure 8.36. The analyzer is set up to sweep from 130 MHz to 160 MHz with a vertical scale of 10 dB/division. The triangle near the top center of the plot indicates the network analyzer marker, which is reading 146.2 MHz and –0.9806 dB. Figure 8.37 shows the phase response of the same filter. Most network analyzers show the phase within a range of ±180 degrees, resulting in the "phase wrapping" that occurs in the middle of the display.

The passband of the filter looks very flat in Figure 8.36. By narrowing the span and expanding the vertical scale, we can zoom in on the passband of the filter (Figure 8.38). With the vertical scale set to 0.1 dB/div, we can see that the filter has less than 0.4 dB of ripple over the frequency span of 144 MHz to 148 MHz.

A measurement using an S-parameter test set results in the display of the forward transfer coefficient of a low-pass filter (Figure 8.39). The resulting measurement indicates S_{21} on the display, which corresponds to the forward gain of the filter. Once again, the vertical scale is log magnitude with a scale of 10 dB/division.

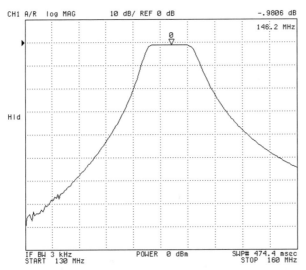

Figure 8.36 The gain through a bandpass filter is measured and displayed in log magnitude format (10 dB per division) using a network analyzer.

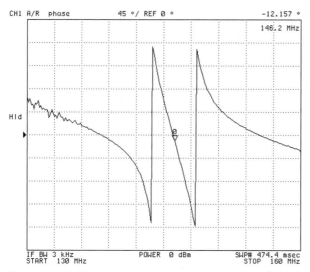

Figure 8.37 The phase response of a bandpass filter is measured using a network analyzer.

Figure 8.38 The flatness of the filter shown in Figure 8.36 is measured by narrowing the frequency span and expanding the vertical scale.

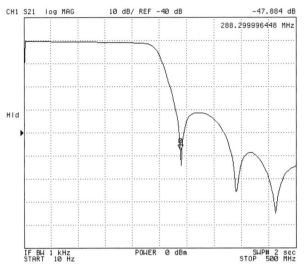

Figure 8.39   The forward transfer coefficient (S$_{21}$) of a low-pass filter is displayed in log magnitude format, corresponding to the gain of the device.

8.9.5   Combination Network/Spectrum Analyzers

We have discussed spectrum analyzers and network analyzers separately. Fundamentally, the basic measurement concepts are distinct, with spectrum analysis describing the content of a *signal* and network analysis describing the characteristics of a *network*. In many cases, the measurement of a signal and its associated circuitry are made side by side. Also, the basic block diagram of a network analyzer is often similar to a spectrum analyzer. As a result, some manufacturers have combined the functionality of a spectrum analyzer and network analyzer in one instrument.

Because of the fundamental differences in measurement concepts, a combined spectrum/network analyzer will usually have two distinct selectable modes corresponding to spectrum measurements and network measurements.

8.10   Distortion Analyzers

Distortion analyzers measure the distortion produced in the circuit being tested. Basically, the distortion analyzer applies a very pure sine wave to the input of a circuit and measures the resulting output. If the circuit is distortion-free, the output will also be a pure sine wave. However, in real circuits, some distortion may be introduced. The distortion meter determines the amount of distortion in the output by filtering out the original sine wave and measuring the RMS voltage of all residual signals, including spurious responses and noise. Since the distortion analyzer is a broadband instrument, high-quality test signals are required for accurate measurements.

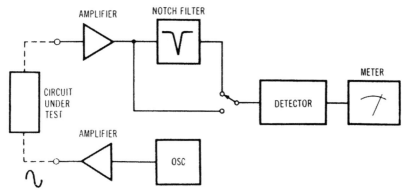

Figure 8.40 The conceptual block diagram of a distortion analyzer. The internal oscillator may be omitted.

Figure 8.40 shows the simplified block diagram of a distortion analyzer. The sine wave output is connected to the circuit under test. (Not all distortion analyzers have internal sources.) This sine wave must be very pure since its distortion will show up as distortion at the output of the circuit and ultimately limit the accuracy of the instrument. The output of the circuit under test is connected to the input of the distortion analyzer. The analyzer uses a narrow notch filter to remove the original sine wave from the input signal. Anything left over is assumed to be distortion in the signal caused by the circuit being tested. The distortion is then compared to the waveform, including the original sine wave, and the distortion is read out on the meter as a percent of the original signal. It is important to note that noise and spurious responses present in the circuit will also contribute to the distortion reading, since they will not be removed by the notch filter. Most distortion analyzers include selectable low-pass and high-pass filtering to help eliminate these unwanted signals.

Simple distortion analyzers only work at one frequency, typically 1 kHz. More advanced distortion analyzers provide the capability of selecting the test frequency. These instruments have a tunable notch filter that automatically tracks the frequency of the internal oscillator. The purity of the measured signal may be expressed in several different ways. For audio purposes, distortion (either in percent or dB relative to the fundamental) is usually preferred:

$$\text{Distortion } (\%) = 100 \, \frac{(\text{Noise} + \text{Distortion})}{\text{Signal}}$$

Notice that noise present in the circuit will add to the measured distortion. Another alternative is expressing the imperfections in terms of either SINAD or SNR (signal-to-noise ratio). These are particularly common in the radio receiver world.

$$\text{SINAD} = \frac{\text{Signal} + \text{Noise} + \text{Distortion}}{\text{Noise} + \text{Distortion}}$$

$$\text{SNR (signal-to-noise ratio)} = \frac{\text{Signal}}{\text{Noise} + \text{Distortion}}$$

8.11 RF Power Measurements

RF power is a fundamental parameter in most radio frequency and microwave systems. For example, communication systems are designed to deliver adequate performance with just enough power in the signal. Poorly regulated signal power generally means using more expensive components and consuming more power supply current. Excessive power can also be a problem, as it may reduce the capacity of a communications system by causing unnecessary interference to adjacent signals.

The *instantaneous power* is defined by multiplying the instantaneous voltage and instantaneous current:

$$p(t) = v(t) \cdot i(t)$$

The *average power* is determined by taking the instantaneous power and averaging it over a period of the waveform.

$$P_{avg} = \frac{1}{T}\int_0^T p(t)dt = \frac{1}{T}\int_0^T v(t) \cdot i(t)dt$$

For practical RF measurements, the averaging takes place over many periods of the waveform. In theory, the averaging time must be an exact multiple of the waveform's period. However, for a large number of periods, the error introduced by averaging over a noninteger number of periods is small.

$$P_{avg} = \frac{1}{NT}\int_0^{NT} v(t) \cdot i(t)dt$$

8.11.1 Pulsed RF

Pulsed RF signals are commonly used in radar and communications systems (Figure 8.41). Figure 8.41a shows a pulsed signal with a pulse width of τ. The *pulse power* is the average of the instantaneous power over the pulse width. Mathematically, it is defined as:

$$P_{pulse} = \frac{1}{\tau}\int_0^{\tau} p(t)dt$$

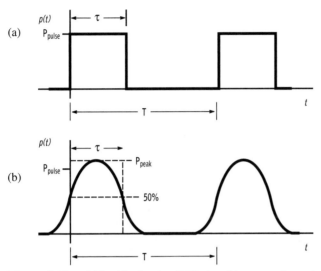

Figure 8.41 a) The ideal pulsed RF signal has a rectangular-shaped pulse. b) A practical pulsed RF signal will have finite rise and fall times.

For a rectangular pulse, pulse power can be related to average power and duty cycle by:

$$P_{pulse} = \frac{P_{avg}}{Duty\ Cycle}$$

In many cases, it is easier to measure the average power in a pulsed signal than it is to measure the actual pulse power. If the duty cycle is known, the pulse power can be calculated from the average power. However, this calculation does depend on having a well-behaved rectangular pulse shape.

Many practical pulsed RF signals do not have a pure rectangular pulse shape. Figure 8.41b shows such a pulsed RF signal. The power at the peak of the signal is called the *peak envelope power* or simply *peak power*.

Peak power, average power, and pulse power are all equal for a pure sine wave.

8.12 RF Power Meter

An instrumentation-grade power meter has an external power sensor that converts the RF signal power to a low-frequency or DC signal, which is read by the power meter (Figure 8.42). The sensor acts as a Z_0 load (50 Ω or 75 Ω) to the device under test. However, it usually cannot handle high power levels (i.e., more than about 100 mW) without an RF attenuator. For power levels up to 25 W, the attenuator may be an integral part of the sensor.

Most power meters measure average power in a signal, while more advanced meters also have the ability to measure peak power. Power level is usually indicated in units of Watts

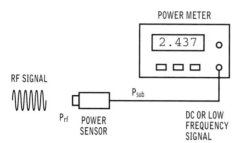

Figure 8.42   An RF power meter uses an external power sensor that converts the RF power to a DC or low-frequency signal for the meter to measure.

Figure 8.43   An RF power meter with external power sensors. (Photo: Agilent Technologies. Reprinted with permission.)

or dBm. While most meters have a single measurement channel, some meters can measure the power level of two signals simultaneously. These dual-channel meters may also have the ability to compute and display the ratio and the difference of the power in the two channels.

Specifications of a typical RF power meter are shown in Table 8.5.

Table 8.5 Abbreviated Specifications of an RF Power Meter

Specification	Value
Frequency range	9 kHz to 110 GHz, sensor dependent
Power range	–70 dBm to +44 dBm (100 pW to 25 W), sensor dependent
Single sensor dynamic range	90 dB
Display units	Watts or dBm
Accuracy (instrumentation)	±0.5%
Accuracy (sensor)	±3%
Accuracy (power reference)	±1.2%

8.12.1 Sensors

The power sensor is a critical piece of technology for power measurements. The power level and frequency of signals that can be measured varies with the particular sensor design, so the user must pay attention to the specifications of the particular sensor being used. A user may need to have multiple power sensors to cover all power measurement needs.

There are three basic types of power sensors, based on their measurement technology: *thermistor, thermocouple,* and *diode detector.* Thermistor and thermocouple sensors both work on the principle of converting the RF signal to heat and measuring the heating effect. This heat-based approach means that these sensors inherently measure the average power of the signal. With today's technology, thermistor sensors have been largely replaced by thermocouple sensors due to their better sensitivity. Diode detectors convert the RF signal into a DC signal that corresponds to the power level in the signal using very fast, precision diodes. The fast response of diode detectors allows them to track signals that are changing, such as pulsed and modulated signals. Modern diode sensors have a dynamic range of 90 dB, able to measure signals ranging in level from –70 dBm to +20 dBm.

There is often amplification and other conditioning and control circuitry in the sensor. The signal from the power sensor is inherently a very small DC signal that varies in voltage proportional to the power measured at the sensor. For a thermocouple sensor, 1 μW of applied power results in approximately a 160-nV signal level. Since it can be difficult to maintain precise DC calibration throughout the signal path, this DC signal is sometimes "chopped" to create a low-frequency AC signal that is easier to handle. A typical chop frequency is 220 Hz.

8.12.2 Calibration Factor

To achieve the best possible accuracy, variations in the measurement of a particular sensor are determined at the factory and correction data is supplied with the sensor. The *calibration factor* for the sensor includes two important sources of error in the sensor: *imped-*

ance mismatch and *sensor efficiency.* Impedance mismatch errors occur when the impedance of the sensor is not exactly Z_0, which causes some of the incident signal to be reflected and not measured.[5] In older power meters, the calibration factor data was printed on the side of the sensor or supplied in the documentation. The calibration factor had to be manually entered by the user, based on the frequency being measured. Advances in digital memory technology now allow the calibration factor data to be stored in nonvolatile memory in the sensor, so that the meter can automatically read the information from the sensor.

Other calibration data may be included with the sensor, such as the measurement error introduced by changes in input power, temperature, and signal frequency. Modern power meter sensors are very complex devices. Typically, they include amplifiers and conditioning circuitry, bias circuits, temperature compensation, digital control circuits, and calibration factor storage.

8.12.3 Zeroing and Calibrating

To maintain the best accuracy for RF power measurements, the meter should be "zeroed" and "calibrated." Zeroing the meter ensures that the meter reads zero when there is no power applied to the sensor. This takes out any residual reading that might be present. To zero a meter, make sure that the power sensor is not connected to a signal source and push the zero control on the meter.

Power meters usually provide a built-in, accurate reference source to calibrate the power sensor being used. This reference is typically a 1 mW (0 dBm) 50-MHz signal. To perform the calibration, connect the power sensor to the reference output and push the "Calibrate" control. The meter will automatically adjust its reading to agree with the reference source.

The zero and calibrate operations just described are intended for the instrument user to optimize the accuracy of the instrument. It should not be confused with the need to periodically send the power meter to a standards lab for a full calibration and adjustment.

8.13   References

"8712E Series RF Vector Network Analyzers—Technical Specifications," Agilent Technologies, Publication No. 5967-6314E, 2000.

Adam, Stephen F. *Microwave Theory and Applications*, Englewood Cliffs, NJ: Prentice-Hall, 1969.

Engelson, Morris. *Modern Spectrum Analyzer Theory and Applications*, Dedham, MA: Artech House, 1984.

---

5.   The calibration factor does not take into account all errors due to impedance mismatch. Some significant errors depend on the impedance of the signal source, which depends on the device being tested. For a detailed discussion of these errors, see "Fundamentals of RF and Microwave Power Measurements," Agilent Technologies.

"EPM-P Series Peak and Average Power Meters User's Guide," Agilent Technologies, Publication No. E4416-90023, June 2001.

"Fundamentals of RF and Microwave Power Measurements," Agilent Technologies, Application Note AN 64-1C, Publication No. 5965-6630E, April 2001.

Gonzalez, Guillermo. *Microwave Transistor Amplifiers*, 2nd Ed., Englewood Cliffs, NJ: Prentice-Hall, 1996.

Oliver, Bernard M., and John M. Cage. *Electronic Measurements and Instrumentation*, New York: McGraw-Hill, 1971.

"PSA Series Spectrum Analyzers—Data Sheet," Agilent Technologies, Publication No. 5980-1283E, September 2001.

"S-Parameter Design," Agilent Technologies, Application Note AN 154, Publication No. 5952-1087, November 2000.

"Understanding the Fundamental Principles of Vector Network Analysis," Agilent Technologies, Application Note AN 1287-1, 2000.

Witte, Robert A. *Spectrum and Network Measurements*, Atlanta, GA: Noble, 2001.

CHAPTER 9

Logic Analyzers

Measurements of digital logic signals can be made with basic test instruments such as an oscilloscope or even a voltmeter. However, there are specialized instruments that can make logic measurements more convenient. The *logic probe* is used to measure individual logic signals, while the *logic analyzer* measures many digital signals simultaneously. Logic analyzers display digital signals as timing waveforms and/or logic state listings. Besides displaying the logic state of the signals, the logic analyzer functions as a data interpreter by displaying the captured logic signals in a variety of forms including decimal numbers, hexadecimal numbers, and microprocessor instruction codes.

9.1 Logic Probes

A logic probe is a small handheld instrument built into a probe-sized case for checking out digital circuits (Figure 9.1). Generally, a logic probe indicates whether the voltage present corresponds to a logic HIGH or a logic LOW. The actual voltage of the signal is not displayed, since it is not significant except whether it's above or below the logic thresholds. Better probes also provide a pulse indicator, which flashes if logic pulses are present. Detailed information such as duty cycle or period cannot be accurately determined with a logic probe, but the presence or absence of a stream of bits can easily be found. The specifications of a typical logic probe are shown in Table 9.1.

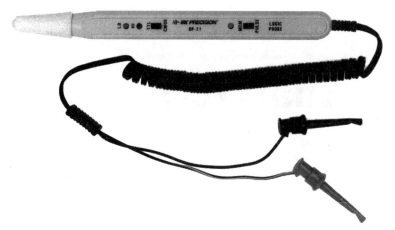

Figure 9.1   A typical logic probe. (Photo courtesy of B&K-PRECISION.)

Table 9.1   Specifications of a Typical Logic Probe

Specification	Value
Frequency range	0 to 50 MHz
Minimum detectable pulse width	10 nsec
Input impedance	2 MΩ
Logic thresholds (TTL)	High = 2.4 V Low = 0.8 V
Logic thresholds (CMOS)	High = 70% of supply voltage Low = 30% of supply voltage

9.1.1   Logic Thresholds

A logic probe is designed to use the appropriate logic levels for the digital technology being tested. Refer back to Chapter 1 (Table 1.6) for details on logic thresholds for various digital logic technologies. The two most common digital logic technologies are TTL and CMOS, which may have different logic thresholds, depending on the particular variant of the technology used. To accommodate both technologies many logic probes may be switchable between TTL and CMOS. Regardless of the type of probe, it is important that the probe's logic thresholds match the logic family being tested.

9.1.2   Logic Probe Indicators

Logic probes indicate the logic state of the digital signal being measured. All logic probes indicate a logic high and a logic low, but other indicators may also be provided. The following list is roughly in order of most common to least common.

Logic High—Indicates that the voltage is greater than high logic threshold, therefore it is a valid HIGH logic signal.

Logic Low—Indicates that the voltage is less than the low logic threshold, therefore it is a valid LOW logic signal.

High Impedance State—Indicates that the voltage is neither a logic low nor a logic high. Normally, this means that the digital gate is in the high impedance state (tri-state) or that the logic probe is not connected to the output of the gate (open circuit). The high impedance state may be indicated by the absence of both the logic high and the logic low indicators.

Pulse—Indicates that the voltage is changing from a valid low logic level to a valid high logic level (or vice versa). Often the regular logic low and logic high indicators are flashed on and off when a pulse occurs. The better logic probes provide pulse stretching circuitry, which catches very short pulses and lights the indicator long enough for the human eye to detect it. Without this special circuitry, short pulses might go undetected by the user, even though the logic level indicator was on for a brief time.

Pulse Memory (or Pulse Trap)—Indicates that a pulse has occurred since the last time the memory was cleared. This is very useful for cases where a digital pulse comes along infrequently but must still be detected. The probe is connected to the circuit and the clear button is pressed, clearing the pulse memory. The indicator stays off until a pulse comes along, then the indicator comes on and stays on until cleared again.

9.2   Oscilloscope Logic Measurements

As discussed in Chapter 5, the oscilloscope is a useful measuring tool for both analog and digital circuits. At first glance, it might seem that a logic probe would not be needed for digital measurements if an oscilloscope is available. This is somewhat true, but the logic probe does have some advantages over the oscilloscope. The oscilloscope gives more information than is required for troubleshooting a digital circuit. Having the voltage waveform displayed, including all of its imperfections, may make it more difficult to troubleshoot a logic circuit. With a logic probe, the user is sheltered from all of the analog information but is given the digital information. The user does not have to remember the logic thresholds of the particular technology being used, but instead just looks at the indicator on the probe to determine the logic state. Of course, if the analog behavior of the circuit is in question (such as too much overshoot or glitches), then an oscilloscope is the right tool.

The other advantage that the logic probe has over the oscilloscope is when very short pulses need to be detected. If a scope is used, the triggering and sweep controls must be adjusted to display the pulse properly. This may be tricky, particularly if the pulse occurs infrequently. If the scope does not have display storage capability, then when the pulse does finally occur it will be a brief flash across the screen.[1] With a logic probe, the logic levels are already set up, so there is no triggering problem. The pulse indicator will detect pulses that are extremely short and display them long enough for a human to see the indicator. If that is not sufficient, then a probe with pulse memory capability can be used to catch the pulse.

9.3   Logic Analyzers

A logic probe provides a convenient means of viewing the state of a logic signal. Unfortunately, a logic probe can view only one signal at a time and, except for the pulse capture feature, can't store a record of the logic signal over time. A *logic analyzer* overcomes these objections by replicating the logic sensing feature of a logic probe, allowing many channels to be measured. A logic analyzer also stores the state of the logic inputs in memory so that a historical record can be displayed.

Since a conventional logic analyzer has two main operating modes, it can be thought of as two instruments in one: a *timing analyzer* and a *state analyzer*. Consider the simplified block diagram of a logic analyzer shown in Figure 9.2. The logic signals being measured enter the logic analyzer via probe buffers, which present a high impedance to the circuits under test. The outputs of the probe buffers are clocked into a latch at regular intervals. The latched data is stored in memory and eventually displayed by the microprocessor.

Note that the signal that clocks the latch can come from two different sources. For *state analysis* (also known as *synchronous operation*), an external clock is used. This clock comes from the circuit under test and is often the master clock in the system. Each rising or falling edge of the external clock causes the logic inputs to be stored into memory, producing a listing of logical information occurring at each clock state. The basic idea of state analysis is to capture the logic data in synch with the circuit under test.

For *timing analysis*, we want to monitor the logic signals on a more continuous basis, independent of the system clock. In timing analysis, a clock internal to the logic analyzer clocks the latch continuously. This type of clocking scheme is referred to as *asynchronous operation*, since the internal analyzer clock runs asynchronous to the logic system being measured.

---

1.   Usually at just the time that the user blinks.

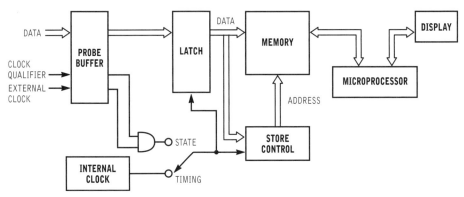

Figure 9.2   The simplified block diagram of a logic analyzer.

Since digital systems can run very fast with clock rates routinely above 50 MHz, it is easy to fill up the memory quickly. The *clock qualifier* input is used to cause the logic analyzer to ignore certain clock cycles that are not useful. Similarly, the *store control* block can be programmed only to store data that meets certain requirements (such as a particular logic pattern) so that memory can be conserved.

The microprocessor reads the data from memory and writes it to the display. The data can be manipulated and displayed in a variety of formats, as discussed later in this chapter.

A typical logic analyzer is shown in Figure 9.3.

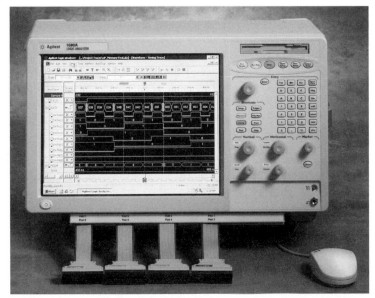

Figure 9.3   A 136-channel logic analyzer. (Photo: Agilent Technologies. Reprinted with permission.)

9.4   Timing Analyzer

When operated as a timing analyzer, the logic analyzer samples the waveform at each edge of its internal clock in a manner similar to a digital oscilloscope. One big difference is that the logic analyzer determines only whether the signal is a logic high or logic low, with no real voltage resolution. Figure 9.4 shows how a voltage waveform is sampled and recorded by a logic analyzer. For each sample interval, the logic analyzer displays the waveform as being high or low.[2]

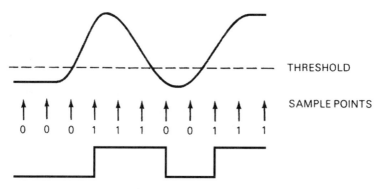

Figure 9.4   The timing analyzer samples a waveform and interprets the voltage as a logic high (1) or a logic low (0).

9.4.1   Timing Resolution

The logic analyzer samples the input signals only on the clock edges and cannot determine what happened in between its internal sample clock edges. This causes a timing uncertainty in the waveform, as shown in Figure 9.5. For a clock frequency of f_{CLK}, the timing resolution (uncertainty) of measurement is $1/(f_{CLK})$. To maximize the timing resolution of a measurement, f_{CLK} needs to be as high as possible. The sample clock may be limited by the speed of the logic analyzer's internal circuitry. But there is a more important tradeoff, which takes place when choosing the clock frequency. The amount of time measured is affected by the clock frequency and the size of the memory.

$$T_{MEAS} = N\, T_{CLK}$$

---

2.   When used as a timing analyzer, a logic analyzer operates similar to a digital oscilloscope but with only one bit of ADC resolution.

where N = the size of the memory

T_{CLK} = the timing resolution ($1/f_{CLK}$)

As f_{CLK} is increased, the timing resolution gets better, but the memory fills up faster.

Example 9.1

What timing resolution is possible given that the total measurement time is 20 msec and the memory size is 100,000?

$T_{MEAS} = NT_{CLK}$, so $T_{CLK} = T_{MEAS}/N = (0.020)/100000 = 20$ nsec.

The available memory can be used more efficiently through the use of *transitional sampling*. With transitional sampling, data is stored into memory only when the input signal changes logic states between samples. Storing the fact that a transition has occurred along with the time since the last transition allows the logic analyzer to completely reconstruct the timing waveform. For signals with frequently occurring transitions, the savings in memory is minimal, but for signals with many samples between transitions (such as data bursts), the amount of memory saved is large. We can think of this memory savings as a way to obtain better timing resolution for some given measurement time, or equivalently, a way to obtain longer measurement time while maintaining a certain timing resolution.

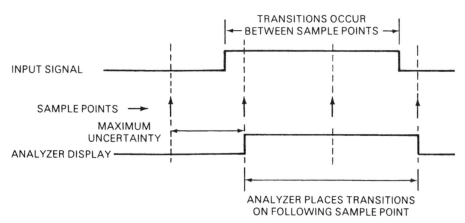

Figure 9.5 The maximum uncertainty in a transition is determined by the sample clock in the timing analyzer.

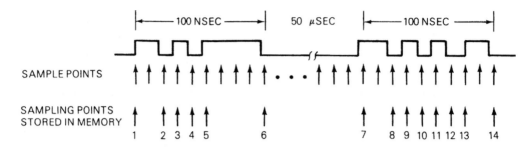

SAMPLE POINTS

SAMPLING POINTS
STORED IN MEMORY

ONLY 28 MEMORY LOCATIONS REQUIRED (14 SAMPLING POINTS + 14 TIME INTERVALS)

Figure 9.6 Transitional timing only stores the transitions into memory, making efficient use of available memory.

Figure 9.6 shows a measurement situation where transitional sampling saves a significant amount of memory without sacrificing timing resolution. The total measurement time is 50 μsec and the desired timing resolution is 10 nsec. Normally, this requires 50 μsec/10 nsec = 5,000 memory locations. However, with transitional sampling, only 14 samples (and their timing information) are stored in memory, requiring a total of 28 memory locations.

9.5   Glitch Detect

Glitches in digital systems are a hardware designer's nightmare and they occur all too often. It is usually difficult to identify glitches and determine their cause. Logic analyzers often provide a special acquisition mode called *glitch capture* or *glitch detect*. In the case of a logic analyzer, a glitch is defined as any transition that crosses the logic threshold more than once between samples (Figure 9.7). Glitches are normally highlighted on the display so they are easy to differentiate from the normal waveform. A dedicated glitch detect circuit is used to monitor the logic analyzer inputs to see if any double (or more) transitions occur between samples. The glitch detect circuit must operate faster than the normal sample rate of the analyzer, but there is a minimum glitch width, which is detectable by the logic analyzer.

The ability to capture and display a glitch helps identify glitches in the timing waveforms. Sometimes a glitch occurs very infrequently, so some logic analyzers provide *glitch*

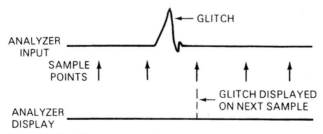

Figure 9.7   Glitches are defined as multiple transitions between samples.

trigger. This feature monitors the analyzer inputs and triggers the analyzer when a glitch occurs. This capability allows the user to set up the analyzer in a "watchdog" role, waiting for a glitch to occur. After the glitch triggers the logic analyzer, the captured waveforms can be analyzed to understand what caused the glitch.

9.6 Digital Logic Test Example

In section 5.5, the operation of a simple 4-bit digital counter is used as an example of an oscilloscope measurement. A logic analyzer can also be used to check the operation of this circuit. Figure 9.8 (same as the Chapter 5 figure) shows the logic signals produced by the 4-bit counter. To display these timing waveforms on a logic analyzer, the four outputs (Q0, Q1, Q2, and Q3) and the clock are connected to the logic analyzer inputs. With the analyzer configured for timing analysis, the counter's clock is treated just like any other logic signal. The logic analyzer display of these signals is shown in Figure 9.9.

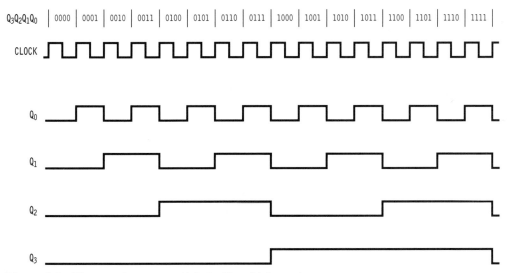

Figure 9.8   The waveforms associated with a 4-bit counter.

Note that this measurement is quite similar to the oscilloscope measurement of Chapter 5. The period and duty cycle of the clock can be checked within the limits of the timing resolution of the measurement. However, the oscilloscope will usually have better timing resolution and will be able to make more precise timing measurements. Another advantage of oscilloscopes are their ability to show the details of the waveform shape, such as overshoot and rise time. Failures due to subtle imperfections in the waveform may not be uncovered by a logic analyzer since the logic analyzer is limited to measuring only logic high and logic low.

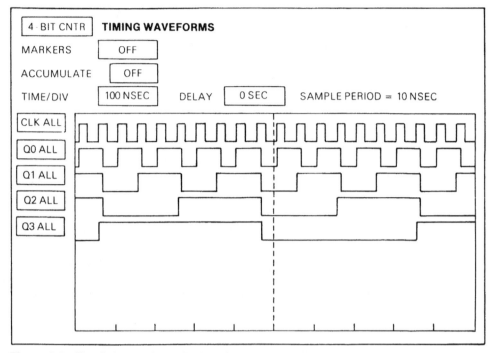

Figure 9.9   The timing analyzer display of the 4-bit counter's waveforms.

On the other hand, most oscilloscopes would not have enough channels to display all five counter signals simultaneously. Our simple 4-bit counter example is instructive, but requires very few channels compared to a typical digital system. The number of channels becomes a much bigger issue in practical digital systems where the address and data buses are often 16, 32, or 64 bits wide. On a complex digital system, the large number of channels available on a logic analyzer are key to capturing and understanding the behavior of the system.

9.7   State Analyzer

When operated as a state analyzer, the logic analyzer uses a signal from the circuit under test as the logic analyzer clock, which determines when the logic signals are stored into memory. Using a clock from the circuit under test means that the logic analyzer runs synchronous with the circuit and acquires the data at just the right time. It also means that the logic signals will not be measured in between these clock pulses, regardless of how the logic analyzer is configured. Compare this to timing analysis, where the logic analyzer's internal clock can be changed to view different amounts of time duration and to obtain a suitable timing resolution. Often, the main clock of the digital system is used as the logic analyzer clock, but not always.

A logic analyzer configured for state analysis can be used to check the operation of the 4-bit counter. Since the counter changes state only on a clock edge, we can use the counter's clock signal as the logic analyzer clock. If the logic analyzer is set to clock on the rising edge (and the counter uses the rising edge), the logic analyzer will capture the logic state of the counter just when the clock goes high. This may be a bit confusing since the logic analyzer is attempting to grab the counter state just as that state is changing. Since it takes a small finite time for the counter to change its state, the counter outputs will remain constant during the rising edge of the clock and for a very short time thereafter. This allows the logic analyzer to record the clock state, which was present just before the clock occurs.

The logic analyzer's display looks much different when doing state analysis. Instead of timing waveforms, the analyzer shows the user a list of logic states, each one acquired on an individual clock edge (Figure 9.10). Note that the state of the clock is not shown because the clock is always at the same state (low) just before a low-to-high transition. The simplest format to use in displaying the logic state is just binary, which shows the state as 1 or 0. For measurement situations using many logic channels, binary patterns can be difficult to interpret and other formats are used.

Measuring the 4-bit counter with both the timing analyzer and state analyzer shows the unique characteristics of each. If the user is interested in checking the timing relation-

4-BIT CNTR	STATE LISTING
MARKERS	OFF

LABEL   >   Q
BASE   >   BIN

+0000	0000
+0001	0001
+0002	0010
+0003	0011
+0004	0100
+0005	0101
+0006	0110
+0007	0111
+0008	1000
+0009	1001
+0010	1010
+0011	1011
+0012	1100
+0013	1101
+0014	1110
+0015	1111

Figure 9.10   The state analyzer display of the 4-bit counter outputs.

ship of the digital signals, timing analysis is the clear choice. On the other hand, if the counter is used in a circuit such that its digital count has meaning, the state analyzer is preferred. The measurement system mirrors the use of the circuit.

9.8   Data Formats

The simplest format for displayed digital signals is binary, with each logic analyzer input being a 1 or a 0. For measurements that have many channels, binary format can become unwieldy. Interpretation of high channel-count measurements can be enhanced by displaying the data in *decimal* or *hexadecimal* (also know as "hex") form.

The decimal form is created from the four binary digits via interpreting each of the bits. From right to left, the binary digits represent the values of 1, 2, 4, and 8. For example, the binary number 0101 has a 1 in the "4 place" and the "1 place." Adding these together and ignoring the zeros, we determine that binary 0101 equals 5.

Hexadecimal form is derived from a base 16 number representation. For values from 0 to 9, hexadecimal is the same as decimal. However, instead of requiring two digits to represent numbers greater than nine, hexadecimal makes use of the letters A through F to represent the decimal values of 10 through 15. Table 9.2 shows the correspondence between binary, hexadecimal, and decimal representations of numbers.

Table 9.2   Binary–Hexadecimal–Decimal Conversion Table

Binary	Hexadecimal	Decimal
0000	0	0
0001	1	1
0010	2	2
0011	3	3
0100	4	4
0101	5	5
0110	6	6
0111	7	7
1000	8	8
1001	9	9
1010	A	10
1011	B	11
1100	C	12
1101	D	13
1110	E	14
1111	F	15

Note: Decimal and hexadecimal are equivalent for values of 9 or less.

Hexadecimal is particularly convenient for representing large binary numbers. For example, the 32-bit binary number 10111000011110001111101000010010 can be divided into 4 bit pieces to obtain 1011, 1000, 0111, 1000, 1111, 1010, 0001, 0010. Hexadecimal representation can be applied to each of these four bits to obtain B878FA12, a more compact form than the original binary.

To facilitate the exchange of alphanumeric information, the Recommended USA Standard Code for Information Interchange (USASCII or, more commonly, ASCII) was developed. This code is used in a wide variety of computers, terminals, and related equipment and is the most common method for encoding textual information into digital form. Most logic analyzers allow the user to interpret and display ASCII encoded data. (The ASCII code is listed in Appendix C.) Besides the usual alphabetic and numeric symbols, ASCII also includes data transmission symbols such as "line feed" and "end of transmission."

Example 9.2

Convert the hexadecimal number A83F to binary form.

Using Table 9.2, we can convert each hex character to binary.
A = 1010, 8 = 1000, 3 = 0011, F = 1111
The binary result is 1010 1000 0011 1111

9.9   State Display

The 4-bit counter represents a simple application of logic analysis that helps us to understand the basic concepts. A state measurement requiring a higher number of channels is shown in Figure 9.11. This measurement consists of 16 bits of status information and a 16-bit address bus. We could view the binary value of each of these channels but as channel count increases this approach becomes very difficult. In the figure, the logic analyzer is set up to treat both Status and Address as *buses*, which means that multiple signals are grouped together and treated as one binary number. The 16 bits of status information is shown as a 4-digit hexadecimal number while the Address bus is shown in both hexadecimal and binary form.

Using state analysis, the logic analyzer clock signal is provided by the device under test. This causes a loss of timing information since the timing of the clock edges is not known. It might be a regularly occurring, stable clock signal or the clock edges may vary considerably, depending on where in the system the clock signal is produced. Most logic analyzers provide time stamps for each state acquired, as shown on the righthand side of Figure 9.11. While usually not as precise as an actual timing measurement, this information provides the logic analyzer user with a useful view of when these states occurred.

Note that the trigger occurred near the middle of the display and that many states were captured that happened before the trigger.

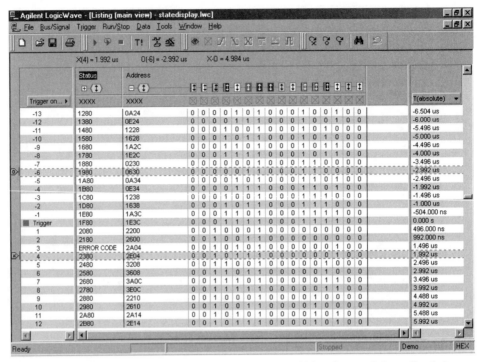

Figure 9.11 This state analyzer measurement shows 16 bits of status information in hexadecimal form and a 16-bit address bus in both hexadecimal and binary forms.

Lists of hexadecimal numbers can be difficult to sort through, so many logic analyzers provide for *symbols* that can be used in place of particular numeric values. For example, in the measurement shown in Figure 9.11, if the Status bits are equal to hexadecimal 2280, it means that an error has occurred in the system. Accordingly, the logic analyzer was set up with a symbol "Error Code" that has the value of 2280. When a status of 2280 is encountered by the analyzer, "Error Code" will be displayed instead of the number. Examination of Figure 9.11 shows that 2280 did occur once during the measurement and is indicated by the "Error Code" symbol.

The symbols feature is a powerful tool that can be used in a number of ways. Symbols are commonly used to indicate important addresses, data locations, error states, reset states, and locations of variables.

9.10 Timing Display

Figure 9.12 shows a timing measurement of a 16-bit data bus and a 4-bit control bus. The individual signals of these buses are displayed along with a consolidated "bus" view of each. When treated as a bus in timing mode, the analyzer overlays all signals in the bus, pro-

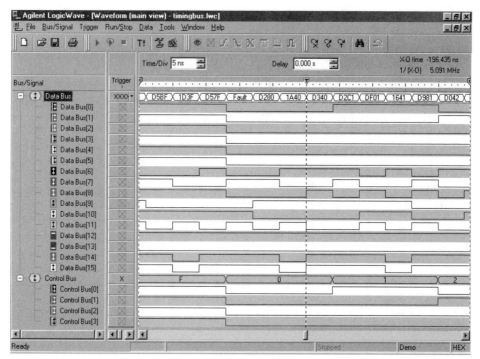

Figure 9.12   This timing analyzer measurement shows a 16-bit data bus and a 4-bit control bus.

viding a composite view of when the bus is changing state. During periods of time that the bus is stable, the analyzer may read out the hexadecimal value that is present on the bus. In Figure 9.12, the data bus is changing faster than the control bus, as indicated by the transitions shown on the two bus waveforms.

The symbol feature discussed with respect to state analysis can also be used with timing analysis. For the data bus, the hexadecimal value of DC80 indicates a fault has occurred in the system, so a corresponding symbol has been defined for the logic analyzer. This symbol can be seen in the data bus display near the top of Figure 9.12.

9.11   Microprocessor Measurement

Choosing an appropriate clock signal for state analysis can be somewhat of an art. The main system clock of a digital circuit is an obvious candidate, but is not always the best choice. Figure 9.13 shows a mythical microprocessor system that includes a main clock, microprocessor, and memory. The timing of the memory read cycle is shown in Figure 9.14. The microprocessor first sets the read/write line (R/W) high to indicate that the next cycle is a read cycle (and not a write cycle). Then the microprocessor outputs the address of the memory location to be read onto the address lines. After these signals have settled to their

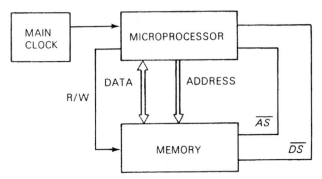

Figure 9.13   A mythical microprocessor system.

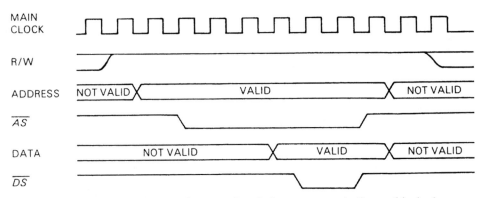

Figure 9.14   The timing diagram for a read cycle from memory in the mythical micropro-cessor system.

valid states, the active-low address strobe $(\overline{AS})^3$ goes low to indicate to the memory that the address is available. The memory responds by producing the data at that address on the data lines. A short time later, the microprocessor pulls the data strobe $(\overline{DS})$ low, then high, clocking the data into the microprocessor.

Suppose there is a problem with reading from the memory. If we need to check the timing of the various signals, the logic analyzer would be configured for timing analysis. But let's suppose that the timing of the circuit is just fine, but the data coming back from the memory is occasionally in error. In this case, state analysis is called for since we just need to monitor the address and data lines for each read cycle. So how do we connect the analyzer to the circuit? Clearly, we need to connect the logic analyzer inputs to the address lines, the data lines, and perhaps the read/write line (so we can tell the difference between

---

3.   The overbar is used here to indicate that the logic signal is Active Low. That is, its main action occurs when the logic line goes into the low state.

read and write cycles). One might be tempted to use the main clock as the clock input to the logic analyzer. Examining the rising edges of the main clock reveals that these edges occur at times when the address and data lines are not valid. So using the main clock as our analyzer clock would result in a state listing with many useless entries. Perhaps the falling edge of address strobe would work since it only occurs once per memory cycle. But no, the data lines are not valid when the falling edge occurs. However, we can clock on the rising edge of address strobe, since both the address and data are valid at that time.

There are likely to be write cycles mixed in with the read cycles of the system. If Address Strobe is used as the clock, the logic analyzer will store both read and write cycles together. To prevent this, the read/write line can be used as a *clock qualifier* with the analyzer set to ignore clocks when the read/write line is in the write (low) state.

9.12   Storage Qualification

Many logic analyzer measurements are an exercise in data reduction. With clock rates in the hundreds of MHz and 64 or more input channels, a logic analyzer is capable of acquiring large amounts of data in a short time. The challenge is sifting through the data to provide meaningful information. Often the logic analyzer's memory fills up before we have captured the desired event. Even when we have enough memory in the analyzer, it may be difficult to sift through the data and find the event of interest.

One feature that can help the user selectively acquire data is called *storage qualification*. Storage qualification allows the user to control the type of data that is stored into the logic analyzer memory. Earlier in the chapter, we discussed using the read/write line to qualify clocks and ignore write operations. We can also use storage qualification to accomplish the same thing by configuring the analyzer to store only logic patterns that have the read/write line high.

Suppose that we discover that the problem with reading the memory in our mythical microprocessor system only occurs within a certain block of addresses. The logic analyzer can be set to store only acquisitions that have the upper address bits set for that particular block of memory. We can also require that the read/write line be HIGH, limiting the stored data to read operations. Alternatively, the read errors may be occurring at all addresses, but only when a certain data pattern is read. If so, the analyzer can be configured to store data on all addresses, but only when the data fits the troublesome data pattern. Using storage qualification lets the troubleshooter narrow the problem down until the faulty logic operation is isolated.

9.13   Trigger Events and Sequencing

Logic analyzers offer an extensive set of advanced triggering capabilities. The advanced triggering capabilities depend on the particular model of logic analyzer, but most analyzers incorporate two basic trigger concepts: *trigger sequencing* and *trigger events*. Sequencing might include constructs such as "Find Event A then trigger on Event B," where the specific trigger events are "Event A" and "Event B."

If we just tell the logic analyzer to run without any particular trigger, it begins storing data on the next clock. Usually, we will want to control when the logic analyzer triggers. Suppose our memory read problem occurs only after memory location 000000 has been read. We could set the logic analyzer to trigger on a read cycle with the address set to 0000000. This means that the analyzer will not trigger until it sees a read cycle from that location, then it will begin storing data according to whatever storage qualification is selected, if any.

Sequencing expands the concept of triggering on a pattern to include finding patterns or other events in a particular sequence. Suppose that our memory read problem is particularly subtle since errors occur in the data only when address 000001 is read, but only after location 000000 is read. To trigger on this sequence, we would set the analyzer to first find a read from address 00000 and then trigger on a read from 00001.

9.13.1 Trigger Events

Here are some trigger events that are typically available in logic analyzers:

Equal to Pattern—Event occurs when the acquired data matches a particular pattern, made up of "1", "0" and "X", where "X" indicates "Don't Care."
Not Equal to Pattern—Event occurs when the acquired data does *not* match a particular pattern.
Greater than Pattern—Event occurs when the acquired data is numerically greater than a specified pattern.
Less than Pattern—Event occurs when the acquired data is numerically less than a specified pattern.
Greater than or Equal to Pattern—Event occurs when the acquired data is numerically greater than or equal to a specified pattern.
Less than or Equal to Pattern—Event occurs when the acquired data is numerically less than or equal to a specified pattern.
In Range—Event occurs when the acquired data is within a specified upper limit and lower limit.
Not in Range—Event occurs when the acquired data is *not* within a specified upper limit and lower limit.

Trigger events can be modified to provide more flexibility.

Time Duration—requires that a pattern or other trigger condition be present for a particular amount of time. For example, trigger if pattern is present for 2 msec.

Count—requires that a pattern or other trigger condition occur a specific number of times. For example, trigger on the 17th occurrence of a pattern.

Advanced logic analyzer triggering systems treat the trigger events listed above as *trigger resources* combining them using logical operators such as AND and OR. For example, a logic analyzer could be set up to trigger when "pattern 1 is equal to 10010XX1" AND "Pattern 2 is less than 01110111." Other trigger resources include counters that can be incremented or reset and flags that can be set or cleared.

Four-way branching using IF/THEN/ELSE controls can be applied to these trigger terms and specific action taken based on this branching. Actions that can result from triggers include *goto* a particular sequence step, *start* or *reset a timer*, *increment a counter*, and *trigger and fill memory*. This type of powerful trigger sequencing looks much like a computer programming language.

Here is an example of a trigger sequence used in an advanced logic analysis system.

```
TRIGGER SEQUENCE

[1] If ADDR In range 000000044 000042A9 Hex And
       Data = XXX03E7 Hex
       occurs 1 time
    then Counter 1 Increment
       Goto 3
    Else if ADDR > 000042A9 Hex
    then Timer 1 Start from reset
       Goto Next
    Else if ADDR < 00000044 Hex
    then Goto 1
[2] FIND EDGE AND PATTERN
    Find *TS Edge ↑
    and ADDR = XXXX43C% Hex
    then Flag 1 Set
         Trigger and fill memory
```

9.14 Microprocessor Program Flow

One of the common applications of logic analyzers is to monitor the program flow in a microprocessor system. The address, data, status, and control lines of the microprocessor are monitored to keep track of what the microprocessor is doing. The processor can be fetching instructions from memory, reading data from memory, writing data to memory, or handling interrupts. Sometimes the logic analyzer is used to debug a problem with the microprocessor software and sometimes it is used to uncover a hardware problem. Sometimes the problem is unknown in origin (hardware or software) or may be a subtle hardware–software interaction. In all of these cases, a logic analyzer can help track down and isolate the problem.

When debugging a microprocessor system, the trigger event and sequencing capabilities are used to track the software program flow. For example, suppose we want to trace the execution of a low-level math subroutine, but only when it is called from a particular section of high-level code. We set the logic analyzer first to find an instruction being executed at an address known to be in the high-level code, then trigger on the entry point of the math routine. The logic analyzer stores only the processor activity starting with the entry to the math routine, saving the analyzer memory for when it's really needed.

The standard probes that come with the logic analyzer can be used to connect to the microprocessor signals, but special adapters called *processor-analysis probes* or *preprocessors* are more convenient. These probes typically plug into the microprocessor socket and the microprocessor plugs into the probe. A processor-analysis probe provides a reliable, fast, and convenient way of connecting to the circuit under test. In addition, it provides clocking and demultiplexing circuits to capture the signals to and from the processor more reliably. Additional status lines may be decoded in the probe, giving the user more information about the internal state of the microprocessor. Analysis probes are also available that handle standard buses and interfaces such as IEEE-488, RS-232C, RS-449, SCSI, VME, and VXI.

Even with a special probe, the address, data, status, and control signals associated with a microprocessor can be difficult to interpret. The logic analyzer user must have a good understanding of the microprocessor's instruction set in order to decode the bit patterns that the logic analyzer displays. Even with this understanding, the logic analyzer user must decode which instruction corresponds to a particular logic pattern. Logic analyzer manufacturers offer *inverse assemblers* (also known as *disassemblers*), which perform this task for the user. The inverse assembler consists of software that runs in the logic analyzer and interprets the instructions and data captured by the analyzer and displays the program flow in the microprocessor's assembly language. The combination of an inverse assembler and processor-analysis probe provides a powerful analysis capability for microprocessor system development. Note that inverse assemblers and processor-analysis probes are designed to work for a particular model of microprocessor. Figure 9.15 shows a typical inverse assembler listing. Shown left to right across the listing are the address of the memory access, the instruction being executed, the operand associated with the instruction, the binary pattern on the data bus, and an indication of whether a read or write is being performed to memory.

M68332EVS	**STATE LISTING**	INVASM
MARKERS	OFF	

LABEL >	ADDR	68010/332 MNEMONIC		DATA _	R/W
BASE >	HEX	HEX		HEX	SYMBOL
+ 0058	6DA4C	MOVE .L	D0, – [A7]	2F00	RD
+ 0059	6DA4E	MOVEQ.L	#00000001, D0	7001	RD
+ 0060	6DA50	ST.B	D0	50C0	RD
+ 0061	02FF4	0000	DATA WRITE	0000	WR
+ 0062	02FF6	0180	DATA WRITE	0180	WR
+ 0063	6DA52	LEA.L	000000, A0	41F8	RD
+ 0064	6DA54	0000	PGM READ	0000	RD
+ 0065	6DA56	LEA.L	4000 [A0], A1	43E8	RD
+ 0066	6DA58	4000	PGM READ	4000	RD
+ 0067	6DA5A	LEA.L	0400 [A0], A5	4BE8	RD
+ 0068	6DA5C	0400	PGM READ	0400	RD
+ 0069	6DA5E	MOVEQ.L	#00000000, D0	7000	RD
+ 0070	6DA60	MOVE.L	[A7] +, D0	201F	RD
+ 0071	6DA62	RTS		4E75	RD
+ 0072	02FF4	0000	DATA READ	0000	RD
+ 0073	02FF6	0180	DATA READ	0180	RD

Figure 9.15   An inverse assembler listing for the Motorola 68332 microprocessor.

9.15   Logic Analyzer Probing

Logic analyzer inputs are connected to the circuit under test using probes provided with the analyzer. Due to the high number of channels that most logic analyzers support, these probes are usually arranged in a pod containing approximately 16 or 32 input channels (Figure 9.16). Like all measurement instruments, logic analyzers disturb the circuit

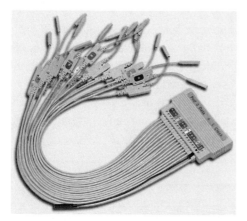

Figure 9.16   Logic analyzer inputs are grouped together into pods. The probe tips shown here attach to the end of a flat cable from the analyzer.

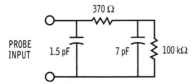

Figure 9.17   A circuit model that represents the loading of a general-purpose logic analyzer input.

under test when the analyzer is attached, so the amount of loading the probe presents to the circuit is important. The concepts of resistive and capacitive loading, discussed in Chapter 4 with respect to oscilloscopes, also apply to logic analyzer probes. A typical logic analyzer probe presents a load of 100 kΩ and 8 pF to the circuit under test (Figure 9.17). 100 kΩ is large enough to avoid significant resistive loading effects on common digital circuits, while the 8 pF of capacitive loading may introduce a small timing error.

Connecting 64 or 128 channels of logic analyzer probes is a time-consuming task. Logic analyzer manufacturers provide a variety of clips and adapters that help with this task. Some digital system designers design in a socket on their circuit board specifically for the purpose of connecting up a logic analyzer. For systems that can support such a connector, such foresight can pay off in the long run via decreased troubleshooting time.

9.15.1   Logic Thresholds

Logic analyzers can usually accommodate a variety of different logic technologies. As discussed in Chapter 1, different logic families have different power supplies and logic levels. The user must specify the logic levels being used and the logic analyzer will set the *logic threshold* appropriately. Alternatively, the user can specify the logic threshold directly.

At the input to the logic analyzer there is a comparator circuit that evaluates the input voltage and decides whether the logic signal is HIGH or LOW. As covered in Chapter 1, there are actually two voltage thresholds that apply to any particular logic level, one threshold that defines the high logic level and one that defines the low logic level. Most logic analyzers employ only a single logic threshold, which is used to determine whether the input signal is high or low. This simplifies the circuitry required in the logic analyzer. More importantly, a logic analyzer is not intended for detailed analysis of the logic waveform. The basic quality of the logic signal and its ability to reach a valid low or high logic state is assumed.

Table 9.3   Abbreviated Specifications of a Logic Analyzer

Specification	Value
Number of channels	68
State clock speed	200 MHz (max)
Timing clock speed	400 MHz (max)

Table 9.3   Abbreviated Specifications of a Logic Analyzer (Continued)

Specification	Value
Memory depth	256 k states (state)
	512 k samples (timing)
Number of State clock qualifiers	4
Trigger sequence levels	16
Maximum input voltage	± 40 V peak
Minimum voltage swing	500 mV$_{P–P}$
Voltage threshold range	–6 V to + 6 V in 10 mV increments

9.16   Combined Scope and Logic Analyzer

Sometimes it is a difficult choice whether to use a scope or a logic analyzer to attack a difficult digital problem. The scope generally provides better timing resolution and a better measure of waveform fidelity while the logic analyzer wins on the number of channels and manipulation of displayed data. Often a system has both analog and digital signals present, so a logic analyzer is insufficient. A scope is not usually up to the measurement task of a complex digital system since it does not have enough channels to capture the digital signals. Instrument manufacturers have solved this dilemma by combining the functionality of a scope and logic analyzer into one instrument.

The scope/analyzer combination produces time-correlated results, which can be displayed separately or simultaneously. Cross-triggering is also available, which allows the pattern trigger and sequencing capability of the logic analyzer to be used to arm or trigger the scope. The logic analyzer can usually be triggered by the scope as well. Complex interactions between digital hardware, software, and analog hardware can be captured with such an instrument.

There are two main approaches to a combination scope/analyzer.

1. **Logic analyzer with oscilloscope channels**—these instruments are full-featured logic analyzers with the addition of two or more oscilloscope channels. These instruments are designed primarily for the scope to be used as an adjunct to the logic analysis feature set. For example, the oscilloscope can be cross-triggered from the logic analyzer when a particular sequence of events occurs. The scope can then be used to view the analog behavior of key critical waveforms.

2. **Mixed signal oscilloscope**—these instruments are full-featured oscilloscopes with timing analyzer channels added that can measure additional digital signals. These instruments are designed to be used primarily as a scope, with the timing analyzer channels providing additional viewing of and triggering on digital signals. Logic

analyzer state analysis is not usually provided. *For more information on mixed signal oscilloscopes, see Section 4.19.*

Figure 9.18 shows a combined analog and digital measurement. The top two waveforms are analog in nature while the bottom four signals are digital timing. The uppermost signal is a series of narrow pulses that would be difficult for a timing analyzer to accurately display. The second signal is an analog waveform that is slowly changing with time. The bottom four signals are digital control signals captured as timing analyzer waveforms.

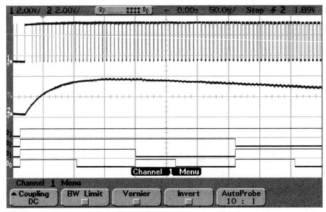

Figure 9.18   A combined analog and digital measurement made using a mixed signal oscilloscope.

9.17   PC-Hosted Logic Analyzer

Since logic analyzers have a large number of channels, their use often results in the collection of copious amounts of data. Analyzing and displaying this data requires a powerful computer system and display to be built into the instrument. Another approach is to make use of a stand-alone PC to handle the analysis and display of data. Figure 9.19 shows a *PC-hosted logic analyzer* that consists of a faceless box containing the logic analyzer acquisition system. The acquisition system is controlled by software that runs on a PC, using graphical user interface techniques.

The PC-hosted design approach has the benefit of being able to use whatever PC happens to be available or fits the needs best. For example, Figure 9.19 shows the use of a notebook PC, which makes the logic analyzer system highly portable. Another option is to provide a PC with a large display to aid in viewing the large number of signals. The PC-hosted approach also tends to be lower in cost, assuming that the user has access to an existing PC.

Figure 9.19   A PC-hosted logic analyzer. (Photo: Agilent Technologies. Reprinted with permission.)

9.18   References

"Agilent 1680 and 1690 Series Logic Analyzers," Agilent Technologies, Publication No. 5988-2675EN, June 2001.

"Agilent Technologies LogicWave E9340A PC-Hosted Logic Analyzer—Technical Data," Agilent Technologies, Publication No. 5968-5560E, November 2000.

Coombs, Clyde F., Jr. *Electronic Instruments Handbook*, 3rd ed., New York: McGraw-Hill, 1999.

"Feeling Comfortable with Logic Analyzers," Agilent Technologies, Application Note 1337, Publication No. 5968-8291E, November 1999.

"Probing Solutions for Agilent Logic Analysis Systems," Agilent Technologies, Publication No. 5968-4632, March 2001.

CHAPTER 10

Circuits for Electronic Measurements

This chapter presents some important circuit concepts that relate to electronic measurements. Electronic circuits are often the devices that are to be characterized or measured. The voltages or currents of some circuit that has already been designed and built often are measured to evaluate the design or to repair the circuit. In other cases, it may be necessary to design and construct a simple circuit to perform the measurement. Circuits that aid electronic measurement may be used to create the measured value. For example, a circuit may be constructed to provide bias current through a diode so that the diode's voltage can be measured. In other situations, the measurement circuit may condition a signal that already exists. For example, an attenuator circuit might be used to reduce the signal level or a low-pass filter might be used to remove unwanted noise.

10.1 Resistance Measurement—Indirect Method

Modern ohmmeters (or multimeters) provide a convenient and accurate means of measuring resistance values. For most applications, this is the fastest and easiest way to make resistance measurements. There are other situations, however, where the ohmmeter is not capable of making the desired resistance measurement. Recall from Chapter 2 that the ohmmeter requires that all power sources be removed from the circuit under test. For measuring devices such as the input and output resistance of an amplifier, this is not possible. In fact, there is no single resistor to be measured since the input and output resistance of an amplifier is the equivalent resistance looking into the input or output. This equivalent resistance depends on the active circuits inside the amplifier. The circuit must have power applied to measure the input or output resistance.

10.1.1 Voltage Divider

One way to measure a resistor is to use an indirect method, which measures circuit parameters other than the unknown resistance and then computes from them the unknown value. Consider our old friend the voltage divider, shown in Figure 10.1. The voltage divider equation for this circuit is

$$V_L = \frac{V_S R_L}{R_S + R_L}$$

Rearranging R_L and R_S can be found in terms of the other values:

$$R_L = \frac{V_L R_S}{V_S - V_L}$$

$$R_S = R_L \left(\frac{V_S}{V_L} - 1 \right)$$

If V_L, V_S, and either R_S or R_L are known, the value of the other resistor can be determined.

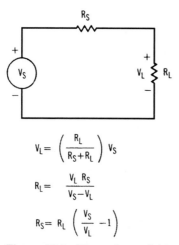

$$V_L = \left(\frac{R_L}{R_S + R_L} \right) V_S$$

$$R_L = \frac{V_L \, R_S}{V_S - V_L}$$

$$R_S = R_L \left(\frac{V_S}{V_L} - 1 \right)$$

Figure 10.1 The voltage divider circuit can be used for measuring the value of an unknown resistance.

10.2 Output Resistance

Suppose the output resistance of a signal source is to be measured. The source's output can be modeled as a Thevenin Equivalent Circuit, as shown in Figure 10.2a. The value of V_S can be determined by measuring the output voltage of the source under open circuit conditions. (Of course, the level of the source depends on its control settings; but for a given instrument setting, V_S will be constant.) A known load resistor can then be connected to the output of the source and V_L can be measured (Figure 10.2b). V_L should always be less than or equal to V_S, due to the loading effect. Some experimentation may be required to determine a suitable value for the load resistor. If R_L is too large, indicated by V_L being very close to or the same as V_S, then the circuit is not being loaded enough. If R_L is too small, indicated by V_L being only a small fraction of V_S, the source may become too heavily loaded and may no longer operate properly. In general, with a suitable value for R_L, V_L should be no more than 90% of V_S and no less than 50% of V_S. Loading the source such that V_L is 50% of V_S eliminates the need for any further calculation since, under this condition, R_S equals R_L.

This technique works on a variety of devices that can be modeled as a voltage source with an internal series resistance. Examples are signal sources, batteries, and (the output of) amplifiers. The same concepts apply whether the voltages involved are DC or AC. In the AC case, it is important to note that the output impedance must be resistive. Fortunately, this is true for many of the output impedances that are measured. As the frequency of the measurement is increased, capacitive and inductive effects become more significant, so this technique is most useful at audio frequencies.

Some devices such as batteries will not tolerate heavy loading and must be measured with as large a load resistor as possible. The reader should also be aware of the current that will flow with a small load resistor. *Large current flow is possible and may damage the device under test or the load resistor.*

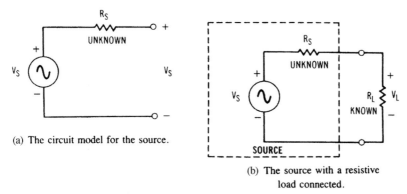

(a) The circuit model for the source.

(b) The source with a resistive
load connected.

Figure 10.2 Using the indirect method to measure the output resistance of a signal source. a) The circuit model for the source. b) The source with a resistive load connected.

Example 10.1

The output resistance of an amplifier is to be determined using the indirect method. A sine wave source was connected to the input, causing a 2-volts RMS AC voltage at the output under no load conditions (Figure 10.3). A 500-Ω load resistor connected to the output caused the output voltage to drop to 1.5 Volts RMS. What is the output resistance of the amplifier?

V_S is equal to the open circuit (no load) voltage, so $V_S = 2$ volts RMS. V_L was measured at 1.5 volts RMS and R_L is 500 Ω. (Note that RMS values are used for V_S and V_L. Zero-to-peak or peak-to-peak could also be used as long as they are used consistently.)

$$R_S = 500\,(2/1.5 - 1) = 167\ \Omega$$

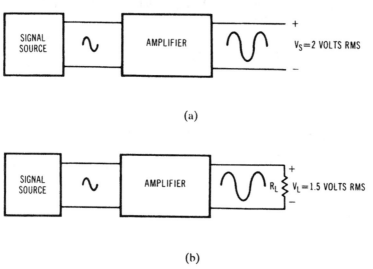

(a)

(b)

Figure 10.3   The output resistance of an amplifier is to be measured. a) A source is connected to the input of the amplifier to produce an AC voltage at the amplifier's output. b) The output amplifier is loaded with a resistor and the output voltage is measured.

10.3   Input Resistance

The same indirect approach can be applied to measuring the input resistance of a device. The input of an amplifier can be modeled as a single resistor, as shown in Figure 10.4a. If a source is connected to such an input, a voltage divider results, with the input resistance of the amplifier acting as the load resistor (Figure 10.4b). With such a connection, neither R_S or R_L can be varied. This is acceptable if a suitable amount of loading occurs (V_L being 50% to 90% of V_S). More often, an additional resistance must be inserted in series

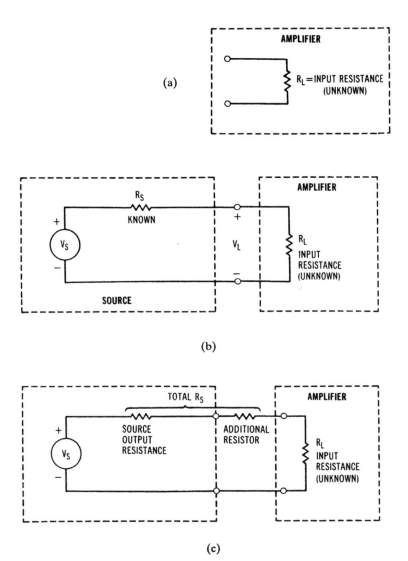

Figure 10.4 The indirect method can be used to determine the input resistance of a device. a) The input of an amplifier is modeled as a resistor. b) A voltage source with nonzero output resistance is connected to the amplifier input. c) It may be necessary to increase the loading effect by adding additional resistance in series with the output of the source.

with the output resistance of the source (Figure 10.4c). This resistor can be chosen, with some experimentation, to provide a reasonable amount of loading. For calculation purposes, the new R_S is the source's output resistance plus the added resistor. The open circuit voltage of the source (V_S) and the loaded voltage (V_L) are measured and R_L is computed. When using an additional series resistor, it is important to measure V_L across R_L (and not across the output of the source). Again, the circuits being measured must be resistive for this method to work properly. The input capacitance of the amplifier can be significant, especially when the input has a very high impedance. Generally, this method is practical for input impedances up to 100 kΩ and for frequencies less than 20 kHz.

Example 10.2

The input resistance of an amplifier is to be measured using the indirect method. A sine wave source with open circuit voltage 0.2 volts zero-to-peak and output resistance 600 ohms is connected to the amplifier using a 10 kΩ resistor (as shown in Figure 10.4c). The voltage at the input of the amplifier is 0.12 volts zero-to-peak. Determine the input resistance of the amplifier.

R_S, for our calculation, is the sum of the source's output resistance and the additional 10 kΩ resistor. R_S = 10.6 kΩ, V_S = 0.2 volts and V_L = 0.12 volts.

$$R_L = \frac{(0.12)\,10.6\text{k}\Omega}{0.2 - 0.12} = 15.9\text{k}\Omega$$

10.4 Bridge Measurements

Another method of measuring resistance and other parameters such as inductance and capacitance is the *bridge circuit*. Resistance measuring methods previously discussed all require accurate voltage and/or current measurements. The bridge method requires only a null indication (minimum voltage reading) and precise reference resistors. Although the instrument user may construct a bridge circuit for measurement use, it is more likely that the user will encounter a commercially made measurement bridge. High-quality ohmmeters have largely replaced resistance bridges, but impedance bridges are still commonly used for measuring complex impedance, inductance, and capacitance. The general concepts concerning bridge measurements will be discussed here to familiarize the reader with the technique.

10.4.1 The Wheatstone Bridge

The *Wheatstone bridge* is the simplest and most easily understood measurement bridge (Figure 10.5). The measurement device (traditionally, a meter) is "bridged" across the circuit. The meter will read zero when the following condition is satisfied:

$$\frac{R_1}{R_2} = \frac{R_4}{R_3}$$

This can be shown by treating the pairs R_1,R_2 and R_3,R_4 as voltage dividers driven by V_S. V_M is the difference between the output voltages of the voltage dividers.

Suppose R_1 equals R_2 and R_3 is the unknown resistor to be measured. The meter will read zero (null indication) when

$$1 = \frac{R_4}{R_3}$$

or

$$R_3 = R_4$$

If R_4 is a variable precision resistor, its value can be adjusted until the meter reads zero. At this setting, the value of the unknown resistor is the same as the value of R_4. This assumes that R_4 has a calibrated dial connected to it so that its value can easily be determined. R_4 may be implemented using a bank of switchable resistors with or without additional variable resistors.

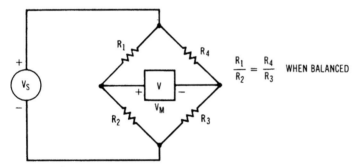

Figure 10.5 The Wheatstone bridge circuit is capable of comparing ratios of resistors very accurately. The meter displays its minimum value when the bridge is properly balanced.

R_1 and R_2 were assumed to be equal in the above discussion. For additional flexibility, these resistors may have other values and may be selectable by the user. The null condition will occur whenever the ratio of R_1 and R_2 equals the ratio of R_3 and R_4. So changing the ratio of R_1 and R_2 alters the ratio of R_3 and R_4, which will produce the null condition. Varying the values of R_1 or R_2 can extend the measurement range of the bridge.

10.4.2 Impedance Bridges

The concept of the Wheatstone bridge can be extended to facilitate impedance measurement. Such a bridge may be used to measure the values of capacitors and inductors. There are quite a variety of bridge configurations that have been developed to accomplish this, but they are all variations on the concept of the resistive bridge. One or more of the resistors in the bridge are either supplemented or replaced by inductors and capacitors. The unknown impedance is one arm of the bridge circuit and one or more of the other arms are adjusted until the null condition is reached. Then the value of the unknown impedance can be inferred from the values of the components in the other arms in the bridge. Of course, since this is an impedance measurement, an AC (sine wave) source is used for V_S.[1]

10.5   RL and RC Circuits

In this section, we explore methods for measuring inductance and capacitance using common lab equipment. These methods are useful for basic measurements and also provide insight into how RL and RC circuits operate. The concepts of step response and time constant can be applied to other areas of electronic circuits.

10.5.1 Step Response

A resistor-capacitor (RC) circuit and a resistor-inductor (RL) circuit are shown in Figure 10.6. These two circuits have some very similar properties that can be used to determine the value of either the inductor or capacitor in the circuit. If the input (V_S) of the circuit is abruptly stepped from 0 volts to some positive value, the output voltage (V_O) rises in an exponential manner (Figure 10.7).

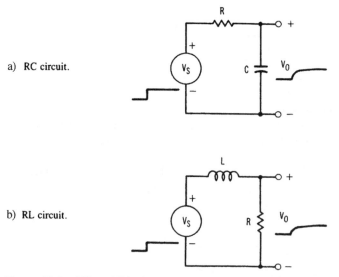

a) RC circuit.

b) RL circuit.

Figure 10.6   RC and RL circuits can be used to determine capacitor and inductor values by measuring the time constant of the circuit. a) RC circuit. b) RL circuit.

---

1.   For a more complete discussion of other bridge circuits, see Helfrick and Cooper (1990) or Jones and Chin (1991).

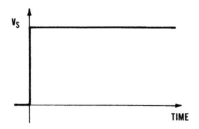

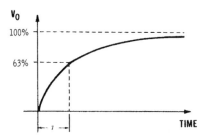

Figure 10.7 The step response input (V_S) and output (V_O) waveforms for the RC and RL circuits shown in Figure 10.6.

The output voltage that results from a voltage step at the input is called the *step response*. The mathematical expression for the step response of the circuits shown is

$$v_O(t) = V_S(1 - e^{-t/\tau}) \qquad \text{for } t \geq 0$$

Notice that the step response does not immediately reach the final V_S level, but rises in an exponential manner. Theoretically, it will take an infinite amount of time for v_O to settle to its final value (which is the same as the final value of V_S). The standard method for describing how quickly the circuit responds is the *time constant* of the circuit.

In one time constant, the step response reaches 63.2% of its final value.

$$v_O(\tau) = V_S(1 - e^{-1}) = 0.632 \cdot V_S$$

The time constant can be computed from the circuit values:

$$\tau = R\,C \qquad \text{for the RC circuit}$$

$$\tau = L/R \qquad \text{for the RL circuit}$$

An unknown capacitor or inductor whose value is to be determined can be connected to a known resistor in the appropriate circuit. (The resistor's value can be measured with an ohmmeter.) The time constant of the circuit can be measured and the unknown component value computed.

$$C = \tau / R \qquad \text{for the RC circuit}$$

$$L = R \, \tau \qquad \text{for the RL circuit}$$

Since the time constant is a time domain parameter, an oscilloscope can be used to measure it (Figure 10.8). A function generator set to output a square wave is an effective way to generate the step input. The function generator is set to output a fairly low frequency (about 100 Hz), which acts as a repetitive step voltage. The period of the square wave must be long enough to allow the circuit to settle within the desired accuracy to its final value before receiving the next voltage step (the rising edge of the next square wave cycle). The amplitude of the square wave is arbitrary, as long as the voltage is large enough to measure easily with the scope, and component ratings are not exceeded.

To a certain extent, the value of the resistor will depend on the value of the capacitor or inductor being measured. At first glance, this may seem like a real problem, since the goal is to measure the capacitor or inductor. However, some experimentation and experience will simplify the process. An experienced user can make a very rough estimate of the size of capacitor by its physical construction. For example, a polarized electrolytic capacitor is likely to be in the 1 μF to 100 μF range, while a small ceramic capacitor is more likely to be less than 1 μF. To use the oscilloscope to measure the time constant conveniently, it is recommended that the time constant be kept within the range of 10 msec to 10 μsec. A recommended resistance value to start with is 1 kΩ. The function generator's output resistance appears in series with the resistance and should be included in the calculations. Figure 10.9 shows the input and output voltages with a suitable time constant and oscilloscope setup.

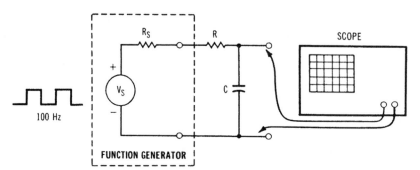

Figure 10.8   A function generator and an oscilloscope are used to measure the step response of this RC circuit. The function generator's output resistance is in series with R and will affect the time constant of the circuit.

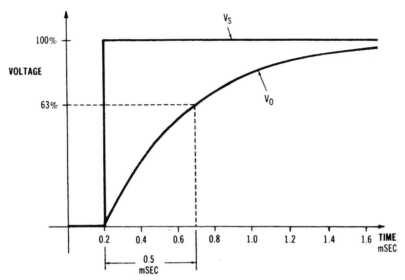

Figure 10.9 An example of an RC circuit's step response. Both input and output waveforms are shown (see Example 10.3).

Example 10.3

The step response shown in Figure 10.9 resulted from an RC circuit with a 1 kΩ resistor and an unknown capacitor. The output impedance of the source is 50 Ω. Determine the capacitor value.

From Figure 10.9, the step response takes 0.5 msec to reach 63% of its final value. Thus, the time constant is 0.5 msec. R is 1 kΩ, but the output resistance of the function generator should also be included (in series).

$$C = \tau/R = 0.5 \text{ msec}/(1000 + 50) = 0.476 \text{ μF}$$

10.5.2 Frequency Response

These same circuits, except with a sine wave source driving them, can be used in the frequency domain to measure the value of a capacitor or inductor (Figure 10.10). The frequency response of these two circuits are the same (Figure 10.11). Low-frequency signals are passed from the input to the output with little or no attenuation, while high-frequency signals are attenuated significantly, resulting in a low-pass filter. Normally, the point at which the response has fallen 3 dB (relative to the response at very low frequencies) is used to define the filter bandwidth. A loss of 3 dB corresponds to a reduction in output voltage to 70.7% of the original value.

$$f_{3dB} = \frac{1}{2\pi\tau}$$

where τ is the time constant of the circuit previously defined. (This is another example of how the time domain and frequency domain concepts are related.)

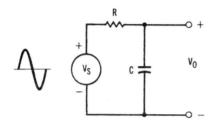

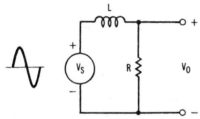

Figure 10.10 The frequency response of the RC and RL circuits can be used to measure a capacitor or inductor.

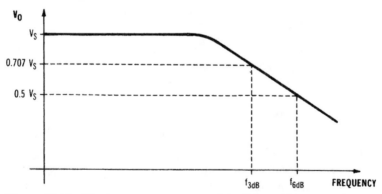

Figure 10.11 The frequency response of the RC and RL circuits shown in Figure 10.10.

Although the 3-dB point is the classical way of defining where a low-pass filter rolls off, other points such as the 6-dB point may be used. A 6-dB reduction in voltage corresponds to a 50% reduction in voltage. This will be a convenient number for measurement use.

$$f_{6dB} = \frac{\sqrt{3}}{2\pi\tau}$$

The equations relating the 3-dB and 6-dB frequencies to the circuit component values are summarized in Table 10.1.

Table 10.1 Equations For RC and RL Low-Pass Circuits

RC Circuit	RL Circuit
$\tau = R \cdot C$	$\tau = L/R$
$f_{3dB} = \dfrac{1}{2\pi\tau} = \dfrac{1}{2\pi RC}$	$f_{3dB} = \dfrac{1}{2\pi\tau} = \dfrac{R}{2\pi L}$
$f_{6dB} = \dfrac{\sqrt{3}}{2\pi\tau} = \dfrac{\sqrt{3}}{2\pi RC}$	$f_{6dB} = \dfrac{\sqrt{3}}{2\pi\tau} = \dfrac{\sqrt{3}R}{2\pi L}$
$C = \dfrac{1}{2\pi R f_{3dB}} = \dfrac{\sqrt{3}}{2\pi R f_{6dB}}$	$L = \dfrac{R}{2\pi f_{3dB}} = \dfrac{\sqrt{3}R}{2\pi f_{6dB}}$

To measure an inductor or capacitor, the unknown component is connected in an RC or RL circuit as appropriate. It is assumed that the resistor is known or can be accurately measured independently. The entire frequency response of the circuit does not need to be measured. It is sufficient to tune the sine wave source to a low frequency (usually 100 Hz or so), note the amplitude of the output voltage using an oscilloscope or meter, and then increase the frequency of the sine wave source until the output voltage drops 3 or 6 dB. The 6-dB point is probably more convenient, occurring when the voltage drops to half of its original value. This technique depends on the output of the source being fairly constant with changes in frequency. (If the source amplitude is not constant with frequency, the frequency response must be calculated at each frequency point as the output voltage divided by the input voltage.) Although the entire frequency response does not have to be measured, it is good practice to trace out the response mentally when adjusting the source frequency. There should be no sharp peaks or dips in the response, just a gradual roll-off with increasing frequency. The 3 dB or 6 dB frequency can usually be determined from the frequency control of the source; or, for more accuracy, a frequency counter can be used to measure the source frequency. After the 3 dB or 6 dB frequency is measured, the value of the inductance or capacitance is calculated using the appropriate equation in Table 10.1.

Example 10.4

The following gain measurements were made on an RL low-pass circuit having a 500-Ω resistor. The output impedance of the source is 600 ohms. Determine the value of the inductor.

Frequency (Hz)	Gain
1000	1.00
2000	0.96
4000	0.87
6000	0.76
8000	0.65
10000	0.57
12000	0.50
14000	0.44

The gain is 0.5 at 12 kHz, so the 6-dB equation is the most convenient in this case. Using the appropriate equation from Table 10.1:

$$L = \frac{\sqrt{3}R}{2\pi f_{6dB}}$$

R in the equation must include both the resistor value and the output impedance of the source. R = 500 + 600 = 1100 Ω.

$$L = \frac{1.732 \times 1100}{2 \times 3.142 \times 12000} = 25.3 \text{mH}$$

10.6   Resonant Circuits

Another property of circuits with inductors and capacitors that can be used for component measurement is *resonance*. At resonance, the impedance of an inductor and capacitor exactly cancel, creating a sharp maximum or minimum in a circuit's response. Consider Figure 10.12a where a parallel LC network is shown. The frequency response of this circuit is a bell-shaped response that peaks at f_0, the resonant frequency. At this frequency, the parallel LC circuit reaches its highest impedance (ideally, infinite). The loading effect of the LC circuit upon the source and resistor is minimized at this frequency. Thus, the output is at a maximum. The resistor value should be set experimentally, starting with about 10 kΩ. A larger resistor causes a sharper resonance (more peaked response) at the expense of a lower output voltage. Generally, the sharper the resonance, the better the measurement.

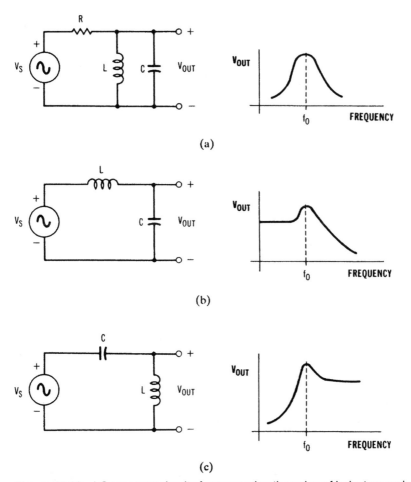

Figure 10.12 LC resonant circuits for measuring the value of inductors and capacitors. a) Parallel LC circuit. b) Series LC with voltage taken across the capacitor. c) Series LC with the voltage taken across the inductor.

The resonant frequency depends only on the L and C values[2]:

$$f_0 = \frac{1}{2\pi\sqrt{LC}}$$

Figure 10.12b and Figure 10.12c show a series version of the LC circuit. Note that no resistor is required in this case. The output voltage is maximum at resonance for both of these circuits, but the shape of the curve is slightly different for each one. The equation for

---

2.   See Appendix B for a more complete analysis of the resonant frequency concept.

the resonant frequency is valid for both series and parallel LC networks. The sharpness of the resonance is dependent mainly on the quality of the inductor and the output impedance of the source (the lower the output impedance, the sharper the resonance).

For measuring either an inductor or a capacitor, the value of the other component in the circuit must be known. The frequency of the source is adjusted until the output voltage peaks—this is the resonant frequency. The value of the unknown component can be calculated using the following equations:

$$L = \frac{1}{4\pi^2 f_0^2 C}$$

$$C = \frac{1}{4\pi^2 f_0^2 L}$$

The choice of which circuit to use depends on the particular circuit components, and is best determined experimentally. The circuit that gives the sharpest resonance is the preferred one. The known component should be chosen to keep the resonant frequency below 10 MHz, where possible. Otherwise, stray capacitance and other parasitic effects will disturb the circuit's response. It is important to be wary of circuit behavior that does not agree with the frequency response curves in Figure 10.12. Multiple resonances are a very common occurrence but can usually be ignored if they are at a much higher frequency than the desired resonance. Responses that do not have a distinct peak are suspect—try another one of the circuit configurations.

Example 10.5

The resonant frequency of the parallel LC circuit (Figure 10.12a) is found to be 18 kHz. The output impedance of the source is 50 Ω and the resistor value is 10 kΩ. If the inductor is 25 μH, what is the value of the capacitor?

Using the equation for the value of the capacitor:

$$C = \frac{1}{4\pi^2 f_0^2 L}$$

$$C = \frac{1}{4\pi^2 (18000)^2 25\mu H} = 3.13\mu F$$

10.7   Diode Measurement Circuit

The forward and reverse voltages of a diode can be tested using a resistor, power supply, and voltmeter. (A more complete characterization of a diode that used an oscilloscope was discussed in Chapter 5.)

The circuit shown in Figure 10.13a can be used to measure the voltage drop across the diode when forward-biased (forward voltage drop). The circuit supplies a positive DC voltage, which forces current through the diode in the forward direction. The voltage across the diode can be measured using a voltmeter or scope. V_S should be set to be larger than the expected forward drop of the diode, but not so large that the diode could be damaged by

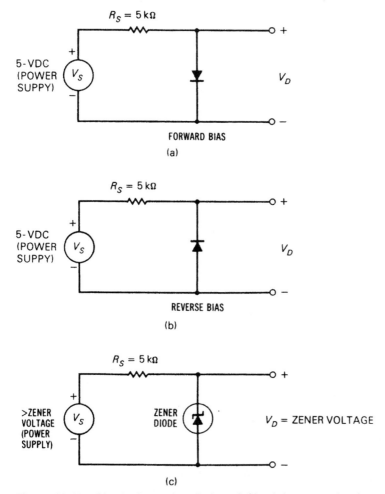

Figure 10.13   Circuits for testing diodes. a) Circuit for measuring forward voltage drop. b) Circuit for testing the reverse bias condition. c) Circuit for measuring the zener voltage of a zener diode.

excessive voltage or current. R_S limits the maximum amount of current that can be forced through the diode. $V_S = 5$ volts and $R_S = 5$ kΩ will forward-bias most common diodes without causing any damage. The forward drop of a functioning diode depends on the type of diode, but is typically less than 1 volt. (Table 10.2 shows the typical values of forward voltage drop of various types of diodes.) A standard silicon diode has a forward drop of about 0.6 volt and a germanium diode has a forward drop of about 0.3 volt. If the measured diode voltage is very close to V_S, then the diode is open circuited and is not functioning properly.

Table 10.2   Typical voltage drops for forward-biased diodes.

Diode Type	Forward Voltage
Germanium	0.25–0.4 volts
Standard silicon	0.60–0.7 volts
Schottky barrier	0.4–0.5 volts
Hot carrier	0.35–0.5 volts
Light emitting (LED)	1.5–2.5 volts

In the reverse direction, the diode should act like an open circuit. This can be verified by reversing the direction of the diode, as shown in Figure 10.13b. The diode voltage should be approximately the same as V_S.

A zener diode will behave like a regular diode until a large reverse bias voltage is applied to it. Under such conditions, the zener diode will limit its voltage (in the reverse direction) at its *zener voltage*. Figure 10.13c shows how the test circuit can be used to measure the zener voltage of a zener diode. V_S must be larger than the zener voltage being measured. Note the polarity of the zener diode—it is being measured in the reverse direction.

10.8   Instrument Connections

Many circuits that are encountered in electronic measurement require that they be connected to a known impedance or resistance for proper operation. This may be the result of having a constant input and output impedance (Z_0) throughout the system, where each piece of the system expects to be loaded by that particular impedance (50 Ω, 75 Ω, etc.). For the measurement to be valid, these same loading requirements must be maintained. One way to accomplish this is to use instruments that are inherently the proper impedance. Another way is to use a *termination*.

10.8.1   Terminations

A termination (or load) is one of the simplest circuits used in electronic instrumentation, but an important one. A termination is nothing more than a resistor, usually packaged

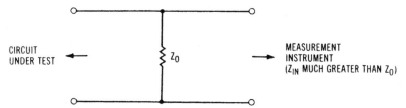

Figure 10.14 A feedthrough termination provides the proper load to the circuit under test, while simultaneously allowing a high impedance instrument to be connected.

with convenient connectors. This resistor can be connected to a circuit under test to provide the proper load impedance. If the termination has two separate connection ports, then it is called a *feedthrough termination* (Figure 10.14). This type of termination is intended to be used with one side connected to the circuit under test and the other side connected to the measuring instrument. If the measuring instrument has a high input impedance (compared to the termination) then the circuit under test is essentially loaded by the termination resistance. (Remember, a small resistance in parallel with a very large resistance is roughly equivalent to just the small resistance.)

10.9 Attenuators

An *attenuator* (also called a *pad*) is a circuit used to reduce the amplitude of a signal. An attenuator may be used to reduce the level of the signal being measured so that a sensitive instrument won't be overloaded. Attenuators also provide some amount of isolation between instruments, reducing any measurement interaction. Commercially built attenuators are available, including well-calibrated adjustable attenuators, but it is often more convenient to build one for a specific application.

Three types of attenuators will be discussed:

1. **Voltage divider attenuators**—Attenuators intended to be driven by low impedance sources and loaded by high impedance circuits or instruments. This definition essentially removes all loading effects due to devices external to the attenuator.
2. **Z_0 attenuators**—Attenuators intended to be driven by sources with a known system impedance (Z_0) and loaded by circuits or instruments with the same system impedance (Z_0).
3. **Impedance matching attenuators**—Attenuators intended to be driven by a particular source impedance (Z_1) and loaded by a different impedance (Z_2). These attenuators are used to connect devices requiring different impedances.

10.9.1 Voltage Divider

In the first case, the attenuator is simply a resistive voltage divider. Again, the loading effects external to the voltage divider itself have been removed so the circuit operation is straightforward. Figure 10.15 shows the voltage divider attenuator and its associated design equations. First, the sum of $R_1 + R_2$ must be chosen arbitrarily. This sum $(R_1 + R_2)$ is the resistive load that will be on the source, so it should not be so small as to overload the source (typically, 1 kΩ to 10 kΩ is reasonable). The values for R_1 and R_2 can be calculated from the equations in Figure 10.15. Some typical values for the case where $R_1 + R_2 = 1$ kΩ are tabulated in Table 10.3. If a larger or smaller resistive load is required, the values in Table 10.3 may be multiplied by a scale factor. As long as the same scale factor is used for both R_1 and R_2, the attenuation remains the same.

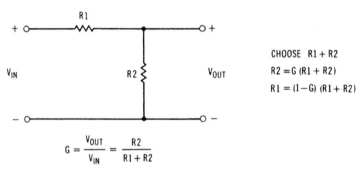

$$G = \frac{V_{OUT}}{V_{IN}} = \frac{R2}{R1 + R2}$$

Figure 10.15 The voltage divider can be used as an attenuator when its output is to be loaded by a high impedance.

Table 10.3 Table of Voltage Divider Attenuator Resistors

Voltage Gain (dB)	Voltage Gain (linear)	R_1	R_2
−1	0.891	109	891
−2	0.794	206	794
−3	0.708	292	708
−4	0.631	369	631
−5	0.562	438	562
−6	0.501	499	501
−10	0.316	684	316
−20	0.100	900	100
−30	0.032	968	32
−40	0.010	990	10

Values shown are for $R_1 + R_2 = 1{,}000$ Ω. These values can be scaled as necessary without changing the gain of the attenuator.

Example 10.6

Determine the resistor values for a voltage divider attenuator whose output voltage is one-tenth of its input voltage. The attenuator should provide no less than a 10 kΩ load to the circuit driving it.

To meet the loading requirement, let $R_1 + R_2 = 10$ kΩ. The voltage gain required in the voltage divider is

$$G = V_{OUT}/V_{IN} = 0.1$$

$$R_2 = G\,(R_1 + R_2) = 0.1\,(10k) = 1\text{ k}\Omega$$

$$R_1 = (1 - G)\,(R_1 + R_2) = 0.9\,(10k) = 9\text{ k}\Omega$$

Alternatively, Table 10.3 shows a 20-dB attenuator (linear voltage gain equals 0.1) with $R_1 = 900$ and $R_2 = 100$. But this is for the case $R_1 + R_2 = 1$ kΩ, which will not meet the loading requirement. These values can be scaled by a factor of 10 to provide an attenuator with $R_1 + R_2 = 10$ kΩ.

$$R_1 = 10 \times 900 = 9\text{ k}\Omega$$

$$R_2 = 10 \times 100 = 1\text{ k}\Omega$$

which is the same answer previously calculated.

10.9.2 Z_0 Attenuators

For systems that have a particular input and output impedance, the attenuator circuits shown in Figure 10.16 can be used. These circuits are designed to be loaded by the same impedance (Z_0) on each end. They are also symmetrical devices—even though an input and output are labeled, they can be reversed in use. For consistency with the voltage divider attenuator, the design equations in Figure 10.16 use voltage gain. Often, attenuators are specified in terms of power gain or loss. Since these particular attenuators have the same impedance at both ends, the voltage gain and power gain are the same *when expressed in decibels*. Figure 10.16a and Figure 10.16b show unbalanced and balanced T attenuators, respectively. Notice that the unbalanced version connects the ground side of the input directly to the ground side of the output. This is not usually a concern if both sides are grounded anyway, but it may be a problem where floating inputs or floating outputs are used (such as on a source, power supply, oscilloscope, etc.). The balanced version of the attenuator solves this problem at the expense of additional resistors. (Note that the R_1 resistors are half the unbalanced value.) Similarly, unbalanced and balanced versions of the Pi attenuator are shown in Figure 10.16c and Figure 10.16d. In operation, the T and Pi attenuator circuits are equivalent. The choice between the two is usually made based on the practicality or convenience of the resistor values.

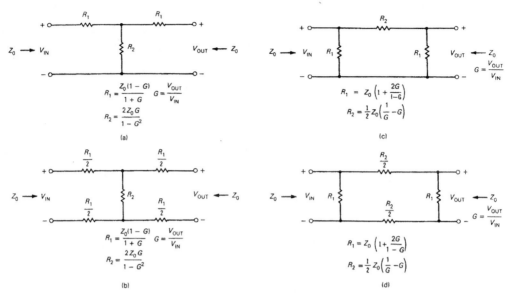

Figure 10.16 Attenuators for use with the same impedance (Z_0) loading each end. a) Unbalanced T attenuator. b) Balanced T attenuator. c) Unbalanced Pi attenuator. d) Balanced Pi attenuator.

Resistor values have been tabulated in Tables 10.4 and 10.5 for attenuators with Z_0 equal to 50 Ω. The values may be scaled to produce attenuators with other Z_0 values. For instance, a 75-Ω attenuator would use resistance values multiplied by 75/50, or 1.5.

Table 10.4 Resistor Values for T Attenuators with Z_0 of 50 Ω

Voltage Gain (dB)	Voltage Gain (linear)	R_1	R_2
−1	0.891	2.87	433.30
−2	0.794	5.73	215.20
−3	0.708	8.55	141.90
−4	0.631	11.31	104.80
−5	0.562	14.00	82.24
−6	0.501	16.61	66.93
−10	0.316	25.97	35.13
−20	0.100	40.90	10.10
−30	0.032	46.93	3.16
−40	0.010	49.00	1.00

These values can be scaled to produce attenuators with Z_0 other than 50 Ω.

Table 10.5 Resistor Values for Pi Attenuators with Z_0 of 50 Ω

Voltage Gain (dB)	Voltage Gain (linear)	R_1	R_2
−1	0.891	869.50	5.77
−2	0.794	436.20	11.61
−3	0.708	292.40	17.61
−4	0.631	220.90	23.84
−5	0.562	178.40	30.39
−6	0.501	150.40	37.35
−10	0.316	96.24	71.15
−20	0.100	61.11	247.50
−30	0.032	53.26	789.70
−40	0.010	51.01	2499.00

These values can be scaled to produce attenuators with Z_0 other than 50 Ω.

Example 10.7

Determine the resistor values for a 10-dB attenuator having both ends loaded by 600 Ω.

We will arbitrarily choose the Pi attenuator shown in Figure 10.16c. The desired loss is 10 dB, which is a gain of −10 dB. Solving for G (not in decibels):

$$G_{dB} = 20\log(G)$$

$$G = 10^{(G_{dB}/20)} = 10^{(-10/20)} = 0.3162$$

$$R_1 = Z_0\left(1 + \frac{2G}{1-G}\right) = 600\left(1 + \frac{2 \times 0.3162}{1 - 0.3162}\right)$$

$$R_1 = 1155\Omega$$

$$R_2 = \frac{1}{2}Z_0\left(\frac{1}{G} - G\right)$$

$$R_2 = \frac{1}{2}(600)\left(\frac{1}{0.3162} - 0.3162\right)$$

$$R_2 = 854\Omega$$

Or using Table 10.4, R_1 = 96.24 Ω and R_2 = 71.15 Ω for Z_0 = 50 Ω. Since the desired Z_0 is 600 Ω, these values must be multiplied by (600/50) = 12.

Example 10.7 (Continued)

$$R_1 = 12 \times 96.24 = 1155\Omega$$

$$R_2 = 12 \times 71.15 = 854\Omega$$

These values can be used for either an unbalanced or balanced Pi attenuator. (If the balanced version is chosen, R_2 is halved.)

10.9.3 Impedance-Matching Attenuators

Although many attenuators are designed and used with the same impedance loading both ends of the device, attenuators can also be designed to work with different impedances at each end. In particular, attenuators can be used to match one impedance to another. Some loss will always be present when using an impedance-matching attenuator, but the loss can be minimized. Attenuators that have the theoretical minimum amount of loss while matching two different impedances are called *minimum loss attenuators* or *minimum loss pads*. Figure 10.17 shows the minimum loss pad circuit diagram. Notice that Z_1 must be the larger of the two impedances. The equations for determining the resistor values and the loss are included in Figure 10.17. Care must be taken in interpreting the loss in dB, since the impedances at each end are not equal.

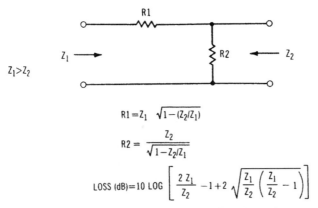

$$R1 = Z_1 \sqrt{1 - (Z_2/Z_1)}$$

$$R2 = \frac{Z_2}{\sqrt{1 - Z_2/Z_1}}$$

$$\text{LOSS (dB)} = 10 \, \text{LOG} \left[\frac{2 Z_1}{Z_2} - 1 + 2 \sqrt{\frac{Z_1}{Z_2} \left(\frac{Z_1}{Z_2} - 1 \right)} \right]$$

Figure 10.17 A minimum loss pad provides impedance matching between two different impedances at the expense of some signal loss. Note that Z_1 must be greater than Z_2.

Minimum loss pads are useful when the measuring instrument input impedance does not match the output impedance of the circuit under test. For example, it may be desirable to measure a 75-Ω system with a 50-Ω instrument input. The 75-Ω system should be loaded by 75 Ω during the measurement and connecting the 50-Ω input directly will result in a mismatch. The problem can be solved by inserting a 50-Ω to 75-Ω minimum loss pad between the circuit under test and the instrument input, thereby providing an appropriate load or match for both the 75-Ω system and the instrument. Of course, there will be some loss in the pad, so the measured values should be adjusted accordingly.

Example 10.8

Calculate the resistor values for a 50-Ω to 75-Ω minimum loss pad. What is the loss of the pad?

Using the equations from Figure 10.17, with $Z_1 = 75\ \Omega$ and $Z_2 = 50\ \Omega$

$$R_1 = Z_1 \sqrt{1 - \frac{Z_2}{Z_1}} = 75\sqrt{1 - \frac{50}{75}} = 43.3\Omega$$

$$R_2 = \frac{Z_2}{\sqrt{1 - \dfrac{Z_2}{Z_1}}} = \frac{50}{\sqrt{1 - \dfrac{50}{75}}} = 86.6\Omega$$

$$\text{Loss(dB)} = 10\log\left[\frac{2Z_1}{Z_2} - 1 + 2\sqrt{\frac{Z_1}{Z_2}\left(\frac{Z_1}{Z_2} - 1\right)}\right]$$

$$= 10\log\left[\frac{2(75)}{(50)} - 1 + 2\sqrt{\frac{75}{50}\left(\frac{75}{50} - 1\right)}\right]$$

$$= 10\log(3.732) = 5.72\text{dB}$$

The resulting circuit is shown in Figure 10.18.

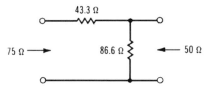

Figure 10.18   A minimum loss pad for connecting a 75-Ω system to a 50-Ω system (see Example 10.8).

10.10   Power Splitters and Combiners

Another termination problem can arise with Z_0 systems when trying to drive two systems from a signal source or when combining two signals into one measuring instrument or circuit. Novice instrument users may attempt to simply connect all of the devices in parallel to combine their outputs. In some cases this may prove satisfactory but usually the impedance matching and loading problems will be significant. A *power splitter* (Figure 10.19) can

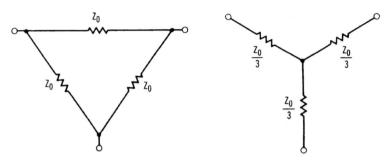

(a) The delta power splitter circuit.   (b) The Y power splitter circuit.

Figure 10.19   These two power splitter circuits are equivalent in operation. a) The delta power splitter circuit. b) The Y power splitter circuit.

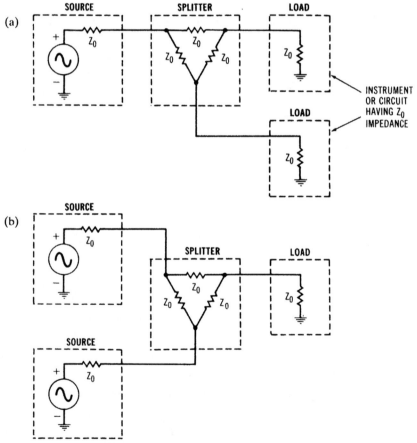

Figure 10.20   The power splitter can be used to split or combine signals. a) Power splitter used for connecting one source to two Z_0 loads. b) Power splitter used for connecting two sources to one Z_0 load.

be used to split one signal into two or to combine two signals into one. The delta power splitter consists of three resistors, each having the value Z_0, connected in a delta arrangement. The Y power splitter also has three resistors but they have the value $Z_0/3$ and are connected in a Y configuration. The two power splitter circuits are totally interchangeable. For proper operation, each connection or "port" of the power splitter must be loaded by a Z_0 impedance. Figure 10.20a shows the delta power splitter in use, with a signal source driving one port and the two other ports connected to Z_0 loads. As long as the power splitter is loaded with Z_0 at each port, the signal level delivered at the two loads is the same. The power splitter can just as easily be used to combine the output of two signal sources into one (Figure 10.20b). Note that each port is once again loaded by Z_0. The power splitter incurs a 6-dB loss (compared with connecting one source directly to one load), which must be accounted for when setting signal source levels or interpreting measured data.

10.11 Measurement Filters

Filters are circuits that pass certain frequencies while rejecting others. In a measurement situation, it may be desirable to insert a filter between the circuit under test and the measuring instrument to remove some unwanted signal or noise that threatens to degrade the measurement. For example, perhaps a low-frequency sine wave is accompanied by high-frequency noise. A low-pass filter could be used to remove the unwanted noise, while allowing the original sine wave to be measured. (This is equivalent to reducing the bandwidth of the measuring instrument.) For general-purpose measurement use, the 3-dB frequency should be chosen far enough away from the desired signal frequency so that it is not attenuated significantly. At the same time, the 3-dB frequency must not be too close to undesired frequency, otherwise that signal will be attenuated enough. Also, keep in mind that for some measurements, such as square wave testing, filtering at any frequency is not appropriate.

Filter design is a complex subject and entire textbooks have been written about it. That level of discussion will not be duplicated here, but a few simple filter circuits can be outlined such that the casual user can make use of them. For more critical applications, it may be necessary to use more complex filter circuits.

Two types of filters will be discussed:

1. **High impedance filters**—Filters intended to be driven by low impedance sources and loaded by high impedance circuits or instruments. This definition essentially removes all loading effects due to devices external to the filter.
2. **Z_0 filters**—Filters intended to be driven by sources with a known system impedance (Z_0) and loaded by instruments with the same system impedance (Z_0).

10.11.1 High Impedance Filters

The two RC circuits shown in Figure 10.21 can be used as filters when followed by high impedance instruments. (The circuit in Figure 10.21a is the same RC circuit used for capacitance measurement earlier in the chapter.) The two circuits share the same 3-dB frequency, but Figure 10.21a has a low-pass frequency response and Figure 10.21b has a high-pass frequency response. The frequency responses of the two circuits are plotted in Figure 10.22. The frequency axis is labeled in terms of the 3-dB frequency, since it is the basic design parameter. Notice that the frequency response does not change abruptly at the 3-dB point, but gradually rolls off.

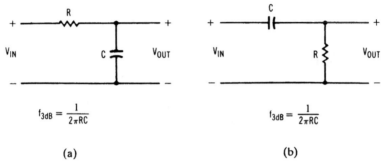

$$f_{3dB} = \frac{1}{2\pi RC}$$

$$f_{3dB} = \frac{1}{2\pi RC}$$

(a)   (b)

Figure 10.21   RC circuits used as filters in high impedance measurement systems. a) The RC low-pass filter. b) The RC high-pass filter.

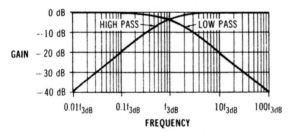

Figure 10.22   The frequency response of the low-pass and high-pass circuits from Figure 10.21. The frequency axis is plotted relative to the 3-dB frequency.

Example 10.9

Design a filter that will attenuate 60-Hz signals coupling from the power line onto the desired 2-MHz sine wave signal. The measuring instrument has a high impedance input.

The design goal is to pass a higher frequency (2 MHz) and attenuate a low frequency (60 Hz). Therefore, a high-pass filter is required. Surveying our extensive collection of high-pass filters, Figure 10.21b is selected. From Figure 10.22, the high-pass characteristic has very little loss at 10 times the 3-dB frequency, so make the 3-dB frequency a factor of 10 less than the desired frequency (2 MHz). Again, looking at Figure 10.22, this should provide a considerable amount of rejection at 60 Hz, which is more than 3,000 times less than the 3-dB frequency.

$$f_{3dB} = 2\text{MHz}/10 = 200\text{kHz}$$

$$f_{3dB} = \frac{1}{2\pi RC}$$

$$RC = \frac{1}{2\pi f_{3dB}} = \frac{1}{2\pi \cdot 200\text{kHz}} = 7.96 \times 10^{-7}$$

Either R or C can be arbitrarily chosen, so choose R = 1000Ω.

$$C = (7.96 \times 10^{-07})/R = 796 \text{ pF}$$

10.11.2 Z_0 Filters

Many measurement applications require a Z_0 termination for filters used to condition the signal. Figure 10.23 shows a low-pass and a high-pass circuit (along with their design equations) appropriate for use in terminated systems. Both circuits are classified as second-order Butterworth networks, a type of filter that has a flat-topped frequency response. The frequency response of the two circuits are shown in Figure 10.24 with the frequency axis shown relative to the 3-dB frequency. The design of such a filter is relatively straightforward—the L and C values are determined by the design equation, depending on Z_0 and f_{3dB}. Both ends of the filter must be terminated in Z_0 for proper circuit operation.

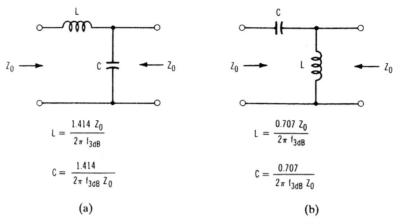

$$L = \frac{1.414\, Z_0}{2\pi\, f_{3dB}}$$

$$C = \frac{1.414}{2\pi\, f_{3dB}\, Z_0}$$

$$L = \frac{0.707\, Z_0}{2\pi\, f_{3dB}}$$

$$C = \frac{0.707}{2\pi\, f_{3dB}\, Z_0}$$

(a)   (b)

Figure 10.23   Filters for use in Z_0 systems. a) Low-pass filter. b) High-pass filter.

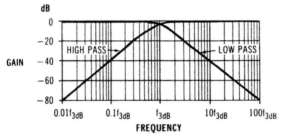

Figure 10.24   The frequency response of the low-pass and high-pass circuits from Figure 10.23. The frequency axis is plotted relative to the 3-dB frequency.

Example 10.10

Design a low-pass filter with a 3-dB frequency of 5 MHz for use in a 50-Ω system.

Using the circuit and equations from Figure 10.23a,

$$L = \frac{1.414 Z_0}{2\pi f_{3dB}} = \frac{1.414 \cdot 50}{2\pi \cdot 5 \times 10^6} = 2.25 \mu H$$

$$C = \frac{1.414}{2\pi f_{3dB} Z_0} = \frac{1.414}{2\pi \cdot 5 \times 10^6 \times 50} = 900 pF$$

10.12 References

Helfrick, Albert D., and William D. Cooper. *Modern Electronic Instrumentation and Measurement Techniques*, Englewood Cliffs, NJ: Prentice Hall, 1990.

Jones, Larry D., and A. Foster Chin. *Electronic Instruments and Measurements* 2nd ed., Englewood Cliffs, NJ: Prentice Hall, 1991.

Schwarz, Steven E., and William G. Oldham. *Electrical Engineering: An Introduction*, 2nd ed., New York: Oxford University Press, 1993.

Williams, Arthur B. *Electronic Filter Design Handbook*, New York: McGraw-Hill, 1981.

Table of Electrical Parameters, Units, and Standard Abbreviations

Quantity	Symbol	Unit	Abbreviation	Comments
charge	Q	coulomb	C	
current	I	ampere	A	ampere = coulomb per second
voltage	V (or E)	volt	V	
resistance	R	ohm	Ω	ohm = volt per ampere
power	P	watt	W	watt = volt-ampere
capacitance	C	farad	F	
inductance	L	henry	H	
frequency	f	hertz	Hz	hertz = cycle per second

Fundamental units (shown) can be modified by using the following standard prefixes:

Prefix	Multiplier	Symbol
femto	10^{-15}	f
pico	10^{-12}	p
nano	10^{-9}	n
micro	10^{-6}	μ
milli	10^{-3}	m
kilo	10^{3}	k
mega	10^{6}	M
giga	10^{9}	G

Examples:

$$2.5 \text{ MHz} = 2.5 \times 10^6 = 2{,}500{,}000 \text{ Hz}$$

$$150 \text{ μF} = 150 \times 10^{-6} = 0.000150 \text{ F}$$

$$25 \text{ mV} = 25 \times 10^{-3} = 0.025 \text{ V}$$

Mathematical Derivations of Equations

B.1   Average (Mean) Value of a Waveform

The average or mean value of a periodic waveform is found by integrating the waveform over one period.

$$\text{average of } x(t) \ = \ \frac{1}{T} \int\limits_{t_0}^{t_0 + T} x(t)dt$$

where

T = the period

t_0 = the start if the integration, which is arbitrarily chosen

B.2   RMS Value of Waveform

The RMS (root-mean-square) value of a waveform is calculated by squaring the waveform, finding its average value, and taking the square root.

$$V_{\mathrm{RMS}} \ = \ \sqrt{\text{average of } v^2(t)}$$

$$= \ \sqrt{\frac{1}{T} \int\limits_{t_0}^{t_0 + T} v^2(t)dt}$$

Example:

$$v(t) = V_{0-P} \sin(2\pi ft)$$

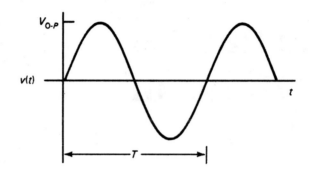

$$V_{RMS} = \sqrt{\frac{1}{T} \int_{t_0}^{t_0+T} [V_{0-P} \sin(2\pi ft)]^2 dt}$$

Choose $t_0 = 0$.

$$V_{RMS} = \sqrt{\frac{1}{T} \int_0^T V_{0-P}^2 \sin^2(2\pi ft) dt}$$

$$= \sqrt{\frac{V_{0-P}^2}{2T} \int_0^T [1 + \sin(4\pi ft)] dt}$$

$$= \sqrt{\frac{V_{0-P}^2}{2T} \left[t - \frac{1}{4\pi f} \cos(4\pi ft) \right] \Big|_0^T}$$

$$= \sqrt{\frac{V_{0-P}^2}{2T} [T]}$$

$$V_{RMS} = \frac{V_{0-P}}{\sqrt{2}} \qquad \text{(sine wave)}$$

B.3   Full-Wave Rectified Average Value of a Waveform

The full-wave rectified average value of a periodic waveform is found by taking the average of the absolute value of the waveform.

$$V_{AVG} = \text{average of } |v(t)|$$

$$= \frac{1}{T} \int_{t_0}^{t_0+T} |v(t)| dt$$

Example:

$$v(t) = V_{0-P} \sin(2\pi ft)$$

Choose $t_0 = 0$.

$$V_{AVG} = \frac{1}{T}\int_0^T |V_{0-P}\sin(2\pi ft)| dt$$

Since $|V_{0-P}\sin 2\pi ft|$ is the same for the time periods of 0 to $\frac{T}{2}$ and $\frac{T}{2}$ to T.

$$V_{AVG} = \frac{2}{T}\int_0^{\frac{T}{2}} |V_{0-P}\sin(2\pi ft)| dt$$

$$= \frac{2V_{0-P}}{T}\int_0^{\frac{T}{2}} \sin(2\pi ft) dt$$

$$= \frac{2V_{0-P}}{T}\left[-\frac{1}{2\pi f}\cos(2\pi ft)\right]\Big|_0^{\frac{T}{2}}$$

$$= \frac{2V_{0-P}}{T}\left(\frac{1}{2\pi f} + \frac{1}{2\pi f}\right)$$

$$= \frac{2V_{0-P}}{\pi fT}$$

since $f = \frac{1}{T}$,

$$V_{AVG} = \frac{2}{\pi}V_{0-P} \qquad \text{(sine wave)}$$

B.4   Bandwidth and Rise Time for a Single-Pole System

A system with a single pole in the frequency domain has a frequency response of the form

$$H(f) = \frac{H_0}{1 + j\left(\dfrac{f}{BW}\right)}$$

where

BW = the 3-dB bandwidth

H_0 = the DC gain

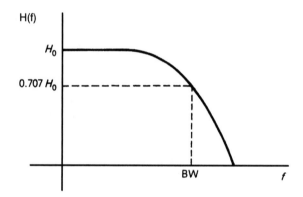

In the time domain, such a system has an exponential step response (assuming $H_0 = 1$)

$$v(t) = 1 - e^{-t/\tau} \qquad t \geq 0$$

where τ is the time constant and

$$\tau = \frac{1}{2\pi BW}$$

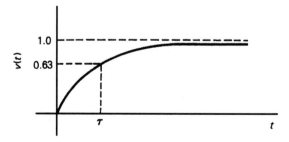

The rise time is defined as the time between the 10% point, t_1, and the 90% point, t_2.

$$0.1 = 1 - e^{-t_1/\tau}$$

$$t_1 = -\tau \ln (0.9)$$

$$t_2 = -\tau \ln (0.1)$$

The rise time, t_r, is

$$t_r = t_2 - t_1 = -\tau(\ln 0.1 - \ln 0.9)$$

$$= 2.197\tau$$

$$= 2.197\left(\frac{1}{2\pi BW}\right)$$

$$t_r = \frac{0.35}{BW}$$

This relationship is exact for a single-pole system and is a good approximation for systems that have very little overshoot in their step responses.

B.5 Frequency Response Due to Coupling Capacitor

The effect of a coupling capacitor can be modeled as

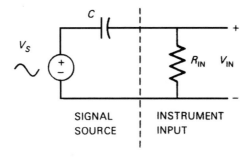

$$V_{IN} = V_S\left(\frac{R_{IN}}{R_{IN} + \frac{1}{j2\pi fC}}\right)$$

$$\frac{V_{IN}}{V_S} = \frac{j2\pi fCR_{IN}}{j2\pi fCR_{IN} + 1}$$

At $f = 0$ (DC),

$$\frac{V_{IN}}{V_S} = 0$$

For large f (high frequency),

$$\frac{V_{IN}}{V_S} = 1$$

At $f = \dfrac{1}{2\pi R_{IN} C}$

$$\left|\frac{V_{IN}}{V_S}\right| = \frac{1}{\sqrt{2}} = 0.707$$

So the 3-dB frequency of this system is

$$f_{3dB} = \frac{1}{2\pi R_{IN} C}$$

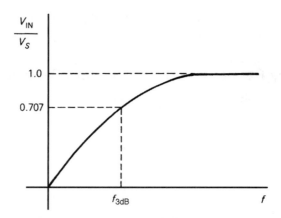

B.6   Polar and Rectangular Formats

A vector number can be represented in both polar and rectangular form.

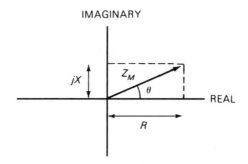

rectangular format: $z = R + jX$
polar format: $z = Z_m \angle \theta$

where

$$Z_m = \sqrt{R^2 + X^2}$$

$$\theta = \tan^{-1}\left(\frac{X}{R}\right) \qquad \text{for } R \geq 0$$

$$\theta = \tan^{-1}\left(\frac{X}{R}\right) + 180 \text{ deg} \qquad \text{for } R < 0, X \geq 0$$

$$\theta = \tan^{-1}\left(\frac{X}{R}\right) - 180 \text{ deg} \qquad \text{for } R < 0, X < 0$$

Other relationships used for converting polar to rectangular format are

$$R = Z_M \cos\theta$$

$$X = Z_M \sin\theta$$

B.7 Resonant Frequency

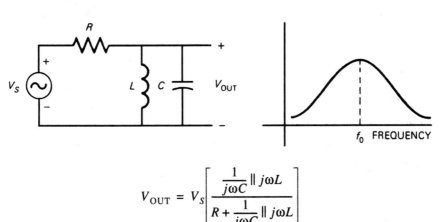

$$V_{OUT} = V_S\left[\frac{\dfrac{1}{j\omega C} \parallel j\omega L}{R + \dfrac{1}{j\omega C} \parallel j\omega L}\right]$$

$$= V_S\left[\frac{\dfrac{(j\omega L)}{(1 - \omega^2 LC)}}{R + \left(\dfrac{j\omega L}{1 - \omega^2 LC}\right)}\right]$$

$$= V_S\left[\frac{j\omega L}{R(1 - \omega^2 LC) + j\omega L}\right]$$

V_{OUT} reaches a maximum when the real part of the denominator is zero:

$$R(1 - \omega_0^2 \, LC) = 0$$

$$\omega_0^2 \, LC = 1$$

Resonance occurs when

$$\omega_0 = \frac{1}{\sqrt{LC}}$$

$$f_0 = \frac{\omega}{2\pi} = \frac{1}{2\pi\sqrt{LC}}$$

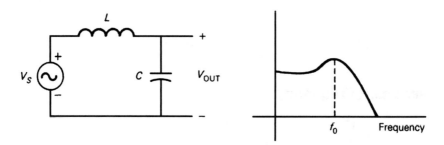

$$V_{OUT} = V_S \left[\frac{\dfrac{1}{j\omega C}}{j\omega L + \dfrac{1}{j\omega C}} \right]$$

$$= V_S \left(\frac{1}{1 - \omega^2 LC} \right)$$

V_{OUT} is maximum when the denominator is zero:

$$1 - \omega_0^2 \, LC = 0$$

Resonance occurs when

$$\omega_0 = \frac{1}{\sqrt{LC}}$$

$$f_0 = \frac{\omega_0}{2\pi} = \frac{1}{2\pi\sqrt{LC}}$$

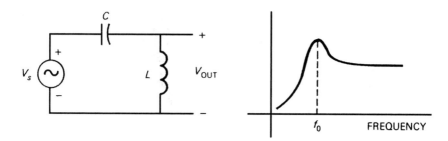

$$V_{OUT} = V_S \left[\frac{j\omega L}{j\omega L + \left(\dfrac{1}{j\omega C}\right)} \right]$$

$$= V_S \left(\frac{-\omega^2 LC}{1 - \omega^2 LC} \right)$$

V_{OUT} is maximum when the denominator is zero:

$$1 - \omega_0^2 LC = 0$$

Resonsance occurs when

$$\omega_0 = \frac{1}{\sqrt{LC}}$$

$$f_0 = \frac{\omega_0}{2\pi} = \frac{1}{2\pi\sqrt{LC}}$$

Binary, Hexadecimal, Decimal, and ASCII Table

Binary	Hexadecimal	Decimal	ASCII	Comments
0000000	00	0	NUL	Null (All Zeros)
0000001	01	1	SOH	Start of Heading
0000010	02	2	STX	Start of Text
0000011	03	3	ETX	End of Text
0000100	04	4	EOT	End of Transmission
0000101	05	5	ENQ	Enquiry
0000110	06	6	ACK	Acknowledge
0000111	07	7	BEL	Bell or Alarm
0001000	08	8	BS	Backspace
0001001	09	9	HT	Horizontal Tabulation
0001010	0A	10	LF	Line Feed
0001011	0B	11	VT	Vertical Tabulation
0001100	0C	12	FF	Form Feed
0001101	0D	13	CR	Carriage Return
0001110	0E	14	SO	Shift Out

Binary	Hexadecimal	Decimal	ASCII	Comments
0001111	0F	15	SI	Shift In
0010000	10	16	DLE	Data link Escape
0010001	11	17	DC1	Device Control 1
0010010	12	18	DC2	Device Control 2
0010011	13	19	DC3	Device Control 3
0010100	14	20	DC4	Device Control 4
0010101	15	21	NAK	Negative Acknowledge
0010110	16	22	SYN	Synchronous Idle
0010111	17	23	ETB	End of Transmission Block
0011000	18	24	CAN	Cancel
0011001	19	25	EM	End of Medium
0011010	1A	26	SUB	Substitute
0011011	1B	27	ESC	Escape
0011100	1C	28	FS	File Separator
0011101	1D	29	GS	Group Separator
0011110	1E	30	RS	Record Separator
0011111	1F	31	US	Unit Separator
0100000	20	32	SP	Space
0100001	21	33	!	
0100010	22	34	"	
0100011	23	35	#	
0100100	24	36	$	
0100101	25	37	%	
0100110	26	38	&	
0100111	27	39	'	
0101000	28	40	(	
0101001	29	41	)	
0101010	2A	42	*	
0101011	2B	43	+	
0101100	2C	44	'	

Binary	Hexadecimal	Decimal	ASCII	Comments
0101101	2D	45	–	
0101110	2E	46	.	
0101111	2F	47	/	
0110000	30	48	0	
0110001	31	49	1	
0110010	32	50	2	
0110011	33	51	3	
0110100	34	52	4	
0110101	35	53	5	
0110110	36	54	6	
0110111	37	55	7	
0111000	38	56	8	
0111001	39	57	9	
0111010	3A	58	:	
0111011	3B	59	;	
0111100	3C	60	<	
0111101	3D	61	=	
0111110	3E	62	>	
0111111	3F	63	?	
1000000	40	64	@	
1000001	41	65	A	
1000010	42	66	B	
1000011	43	67	C	
1000100	44	68	D	
1000101	45	69	E	
1000110	46	70	F	
1000111	47	71	G	
1001000	48	72	H	
1001001	49	73	I	
1001010	4A	74	J	

Binary	Hexadecimal	Decimal	ASCII	Comments
1001011	4B	75	K	
1001100	4C	76	L	
1001101	4D	77	M	
1001110	4E	78	N	
1001111	4F	79	O	
1010000	50	80	P	
1010001	51	81	Q	
1010010	52	82	R	
1010011	53	83	S	
1010100	54	84	T	
1010101	55	85	U	
1010110	56	86	V	
1010111	57	87	W	
1011000	58	88	X	
1011001	59	89	Y	
1011010	5A	90	Z	
1011011	5B	91	[	
1011100	5C	92	\	
1011101	5D	93	]	
1011110	5E	94	^	
1011111	5F	95	–	
1100000	60	96	`	
1100001	61	97	a	
1100010	62	98	b	
1100011	63	99	c	
1100100	64	100	d	
1100101	65	101	e	
1100110	66	102	f	
1100111	67	103	g	
1101000	68	104	h	

Binary	Hexadecimal	Decimal	ASCII	Comments
1101001	69	105	i	
1101010	6A	106	j	
1101011	6B	107	k	
1101100	6C	108	l	
1101101	6D	109	m	
1101110	6E	110	n	
1101111	6F	111	o	
1110000	70	112	p	
1110001	71	113	q	
1110010	72	114	r	
1110011	73	115	s	
1110100	74	116	t	
1110101	75	117	u	
1110110	76	118	v	
1110111	77	119	w	
1111000	78	120	x	
1111001	79	121	y	
1111010	7A	122	z	
1111011	7B	123	{	
1111100	7C	124	|	
1111101	7D	125	}	
1111110	7E	126	~	
1111111	7F	127	DEL	Delete

INDEX

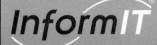
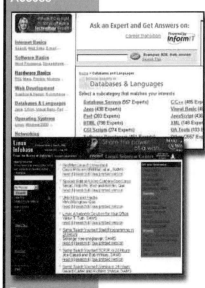